BIODIVERSITY COMMUNITIES AND CLIMATE CHANGE

BIODIVERSITY COMMUNITIES AND CLIMATE CHANGE

Editors
Chandra Prakash Kala
Chandra Shekhar Silori

teri
The Energy and Resources Institute

First reprint 2017

ISBN 978-81-7993-442-5

Suggested citation

Kala, Chandra Prakash, and Chandra Shekhar Silori. 2013. *Biodiversity, Communities, and Climate Change.* New Delhi: TERI

Published by

The Energy and Resources Institute (TERI)
TERI Press
Darbari Seth Block
IHC Complex, Lodhi Road
New Delhi – 110 003
India

Tel. 2468 2100 or 4150 4900
Fax 2468 2144 or 2468 2145
India +91 • Delhi (0)11
E-mail teripress@teri.res.in
Website www.teriin.org

Printed in India

Foreword

Recently, the need for a greater understanding of linkages between biodiversity conservation, climate change mitigation and adaptation, and the livelihoods of local communities has increased the attention of world leaders. The threat of climate change, which cuts across national and international boundaries, has further emphasized the importance of the issue and facilitated conservationists and development practitioners to join hands in finding solutions to restore rapidly vanishing forests and biodiversity. Conservation of biodiversity and associated ecosystem goods and services is important for upholding the economic, social, and cultural values and rights of forest-dependent communities. The limited effectiveness of two extreme approaches that emphasize either exclusive biodiversity conservation (for example, protected area system) or livelihoods based (for example, community development) is evident. More recently, however, the general tendency towards increased recognition of community forestry achievements and the need for a more inclusive and participatory approach for protected area management are steps in the right direction.

Our understanding of certain aspects of biodiversity conservation using inclusive approaches such as community forestry, especially in the context of climate change, is still very limited. *Biodiversity, Communities, and Climate Change* addresses these issues. It draws upon practical work experiences of field-based research on diverse biodiversity related topics such as genetic to ecosystem diversity, use of latest technology (including GIS and remote sensing in forest management), participatory conservation approaches, importance of traditional and indigenous knowledge and its role in biodiversity conservation, and adaptation to climate change. It also focuses on the need for community conservation to move beyond traditionally perceived biodiversity conservation problems and solutions and to engage in new partnerships that help find ways to balance conservation and development needs in rapidly changing social, economic, and environmental contexts.

I would like to congratulate the editors and authors for taking the time and effort to compile practical experiences from the field and making a contribution to enhance our understanding of issues facing the conservation of biodiversity and their possible solutions. This book will serve as a useful source of information for students, researchers, development planners, and policymakers who are concerned about the conservation of biodiversity and the threats of climate change.

Yam Malla, PhD
Executive Director
RECOFTC-The Center for People and Forests,
Bangkok, Thailand

Preface

For decades, biodiversity conservation and the sustainable use of natural resources have been the key challenges for development agencies and concerned stakeholders alike. The looming threat of climate change has only added to this complexity. Although evolution and the extinction of species are considered natural phenomena, the evidence of previous mass extinctions points mainly to massive climatic or environmental shifts. At present, the alarming rate of species extinction is accelerated by human activities for various developmental and livelihood needs and aspirations.

Documented evidence indicates that the loss of even a single species can have severe impacts on an ecosystem. Despite this fact, there is inadequate information and knowledge about the existing flora and fauna, which increases the challenge of conservation. A number of studies on biodiversity and its conservation exist but the challenge continues to be the striking of a balance among conservation, livelihood security, and countering the potential impacts of climate change.

In this context, this book, *Biodiversity, Communities, and Climate Change*, is designed to help in guiding the various principles of biodiversity conservation, effects of climate change, and the role of communities at various levels and in different landscapes. The book draws on the experiences of academicians and eminent scientists in the form of case studies and reviews revolving around key themes. There are 19 chapters, which cover a wide range of topics including tools of biodiversity assessment, ranging from conventional, ecological, and social survey methods to the use of the latest technology such as Geographical Information System (GIS) and remote sensing.

The topical issues in different chapters range from local to regional levels and cover concerns for communities in livelihood security and poverty alleviation, conservation, effects of climate change on biodiversity and communities, and possible solutions for conserving the biodiversity and the ecosystem. The chapters are arranged under three broad thematic categories.

The first eight chapters focus on the theme of biodiversity conservation. These describe the status of biodiversity from lower to higher flora and fauna and various assessment methods, including the applications of remote sensing and GIS in mapping plant resources and evaluating the impact of allelopathic compounds on the growth of native plant species, and geospatial techniques for monitoring urban ecosystem and biodiversity.

The second section of the book is a compilation of chapters dealing with community, livelihood, and the management of biodiversity. The seven chapters cover participatory conservation and forest management case studies and key lessons learnt to further strengthen conservation approaches.

The third section of the book has four chapters that focus on the potential roles of community forestry, the agriculture sector and biodiversity conservation, including cultivation of medicinal plants and trees in addressing climate change threats.

The first chapter, by Sharma and Pati, experts of biodiversity, examines the fungal diversity in tropical forests and their possible effects on host plant species. In Chapter 2, entomologists Quasin and Uniyal report on the status and composition of spider diversity along an altitudinal gradient in a well-known protected area, the Nanda Devi Biosphere Reserve of the Indian Himalayas. The authors of Chapter 3 are renowned botanists Kala and Majila who explain in detail the species diversity, composition, dominance, natality, and mortality of woody plant species in the Binsar Wildlife Sanctuary of Uttarakhand. The authors also analyse the status of wildlife and the issues of man–animal conflict in and around the sanctuary. In Chapter 4, botanist and medicinal plant expert, Kala talks about the social and ecological dimensions of medicinal plants and their traditional harvesting and management practices in the deciduous forests of India.

Chapter 5 by Pattanaik, Reddy, and Roy, specialists of remote sensing in landscape conservation, examine the application of remote sensing and GIS in mapping biological richness by using a spatial model incorporating ground-based biodiversity attributes of the landscape elements, land use change patterns, disturbance regimes of the landscape, and terrain complexity.

The authors of Chapter 6, landscape conservationists Yousaf and Haq report on the chemical relationship of an invasive plant species, *Parthenium hysterophorus*, with the native plants. Chapter 7 by plant geneticists Khanum and Mumtaz, assesses the role of genetic factors in contributing vulnerability in plants and in shaping the plant population. In Chapter 8, landscape conservationists Shariff and Gairola review the potential need for urban biodiversity studies using geospatial tools for sustainable management.

The role of communities is always a matter of concern, not only in exploring various uses of biodiversity but also in developing sustainable utilization techniques and customary norms. Being a source of their livelihood, traditional ecological knowledge and practices associated with biodiversity have proven to be an advantage in safeguarding the biological diversity. The second section of this book, Chapters 9 to 15, therefore, envisage issues concerning communities, livelihood, and management of biodiversity. Chapter 9 by Silori, social ecologist and community forestry expert, discusses a case of self-initiated community conservation of forest recovery and its consequent environmental and social benefits in the Doi Suthep Pui National Park, Thailand. In Chapter 10, medicinal plant conservationist Badoni throws light on livelihood security issues through the cultivation of medicinal plants, especially the threatened plant species *Kuth* in the Uttarakhand Himalayas.

In Chapter 11, Singh and Sundriyal, botanists, report the pattern of the traditional agroecosystem, sustainability of agriculture systems, the status of fuelwood and fodder, cash income from the agriculture system, and different income groups in a Central Himalayan village. In Chapter 12, Silori explains the process of joint management in protected areas and the roles of various actors and the resultant benefits in terms of livelihoods of the dependent communities and the status of biodiversity conservation in Phutoei National Park,

Thailand. Chapter 13 by Kala and Farooquee highlights the links between the indigenous knowledge, the community decision-making process and conservation of natural resources in the high altitudes of the Uttarakhand Himalayas using the example of *Zanthoxylum armatum*.

In Chapter 14, ethnobotanists, Rana, Tiwari, Dangwal and Gaur present a brief account of the ethno-taxonomy of medicinal plants in the Nanda Devi Biosphere Reserve along with the indigenous belief system of the local people in traditional healthcare. The authors suggest that the indigenous belief system provides the opportunity to set up small-scale enterprises in rural areas, particularly aiming to provide livelihood support through various interventions. Chapter 15 by Kala examines the uses of non-timber forest products in high-altitude tribal communities. The author reports that the traditional method of making fermented food and beverages is designed in such a way as to utilize available crops that do not comprise the main food crop.

Worldwide, the unprecedented loss of biodiversity has been attributed mainly to escalating anthropogenic activities, which include an increase in the concentration of greenhouse gases in the atmosphere and consequent changes in the climate. Such a change in the climate has brought different kinds of changes in the structures of forests and biodiversity within them. The last section of this book includes four chapters, which discuss some of such changes in the biodiversity and the community perspective regarding them.

In Chapter 16, Kala reviews the effects of climate change on biodiversity and the challenges in clean development mechanism (CDM) and climate change mitigation strategies. Chapter 17 by Silori examines the role of forests and the associated traditional knowledge of local communities and indigenous people in climate change adaptation. In Chapter 18, Bijalwan, an agro-ecologist, reviews various adaptation and mitigation strategies of climate change in Indian agriculture.

In the final chapter, foresters and biodiversity conservationists Chauhan, Gera and Aggarwal report on climate change mitigation potential and the cost effectiveness of growing tree species of medicinal value on forest lands by using the PRO-COMAP model for the estimation of recurring carbon benefits from mandatory as well as voluntary markets. The authors point out that some medicinal tree species such as *Terminalia arjuna*, *Terminalia belleirica*, and *Terminalia chebula* have more carbon sequestration potential and hence motivate growers to transfer more land for their plantations.

The objective of this book is to incorporate different viewpoints of scholars working on the interdisciplinary issues of biodiversity conservation, climate change, and livelihood security of communities. This book will serve as a useful reference guide to the scholars, academicians, policymakers, scientists, and different stakeholders working on one or more of the themes covered in the book by answering several critical questions and helping them design the future course of action.

List of Contributors

Abdual Samad Mumtaz, Associate Professor, Quiad-i-Azam University, Islamabad 44000, Pakistan. Email asmumtaz@qau.edu.pk

Arun K. Badoni, Director, Society for Himalayan Environmental Research (SHER), Lane No-3, Turner Road, P.O. Clement Town, Dehradun 248002, Uttarakhand, India. Email sher@sancharnet.in

Arvind Bijalwan, Assistant Professor, Technical Forestry, Indian Institute of Forest Management, P.B. No. 357, Nehru Nagar, Bhopal 462 003, Madhya Pradesh, India. Email arvindbij@iifm.ac.in; arvind276@rediffmail.com

Ashish Aggarwal, Ph. D. Scholar, School of Environment and Development, The University of Manchester, UK. Email enviroashish@gmail.com

Bhagwan Singh Majila, Scientist, G.B. Pant Institute of Himalayan Environment & Development, Kosi-Katarmal, Almora 263643, Uttarakhand, India.

C. Sudhakar Reddy, Scientist, National Remote Sensing Centre, Indian Space Research Organization, Hyderabad 500625, Andhra Pradesh, India.

C.S. Rana, Postdoctoral Fellow, Society for Himalayan Environmental Research (SHER), Lane No 3, Turner Road, P.O. Clement Town, Dehradun 248002, Uttarakhand, India. Email drcsir@gmail.com

Chandra Prakash Kala, Faculty, Ecosystem & Environment Management, Indian Institute of Forest Management, Post Box

357, Nehru Nagar, Bhopal 462003, Madhya Pradesh, India. Email cpkala@iifm.ac.in; cpkala@yahoo.co.uk

Chandra Shekhar Silori, Senior Programme Officer, RECOFTC – The Center for People and Forests, Bangkok, Thailand. Email silori_cs@yahoo.com, chandra.silori@recoftc.org

Chiranjibi Pattanaik, Consultant, Environment Division, Research & Development Complex, Engineers India Limited, Sector 16, On NH-8, Gurgaon 122001, Haryana, India. Email chiranjibipattanaik@gmail.com

J.K. Tiwari, Professor, Department of Botany, HNB Garhwal University, Srinagar (Garhwal) 246 174, Uttarakhand, India.

Kavita Sharma, Head, Department of Botany, Arts and Commerce Girls College, Raipur, Chhattisgarh 492001, India. Email: drktsharma@gmail.com

L.R. Dangwal, Assistant Professor, Department of Botany, HNB Garhwal University, Srinagar (Garhwal) 246 174, Uttarakhand, India

Mohit Gera, Member Secretary, State Pollution Control Board, Jammu and Kashmir, India. Email mohitgera87@gmail.com

Narayan Singh, Project Scientist, National Mission on Medicinal Plants, Herbal Research & Development Institute, Mandal, Gopeshwar, Chamoli 246401, Uttrakhand, India. Email naturewithnary@gmail.com

Nehal Ahmad Farooquee, Associate Professor, School of Extension and Development Studies, Indira Gandhi National Open University (IGNOU), Maidan Garhi, New Delhi 110068, India. Email nafarooquee@rediffmail.com

Noresah Mohd Shariff, Associate Professor, Geography Programme, School of Distance Education, Universiti Sains Malaysia, 11800 Minden, Penang, Malaysia. Email noreshah@usm.my

P.S. Roy, Director, National Remote Sensing Centre, Indian Space Research Organization, Hyderabad 500625, Andhra Pradesh, India.

R.C. Sundriyal, Scientist, G.B. Pant Institute of Himalayan Environment & Development, Kosi-Katarmal, Almora 263 643, Uttarakhand, India. Email rcsundriyal@gbpihed.nic.in

R.D. Gaur, Professor Emeritus, Department of Botany, HNB Garhwal University, Srinagar (Garhwal) 246 174, Uttarakhand, India.

R.N. Pati, Executive Director, Council for Tribal and Rural Development (CTRD), Mahamaya Bhawan, HIG-101, Kanan Vihar, Phase-I, P.O: Patia, Bhubaneswar, Odisha, India. Email drpati54@gmail.com

Rizwana Khanum, Ph.D. Scholar. Quiad-i-Azam University, Islamabad 44000, Pakistan. Email rizvana.khan@gmail.com

Sanjay Gairola, Postdoctoral Fellow, School of Agricultural, Earth and Environmental Sciences, College of Agriculture, Engineering and Science, University of KwaZulu-Natal, Durban 4000, South Africa. Email drsanjaygairola@gmail.com

Shazia Quasin, Research Associate, Wildlife Institute of India, Post Box # 18, Chandrabani, Dehradun 248001, Uttarakhand, India.

Suresh Chauhan, Fellow, Forestry and Biodiversity Group, Climate Change Division, The Energy and Resources Institute (TERI), Darbari Seth Block, India Habitat Center, Lodhi Road, New Delhi 110003, India. Email sureshc@teri.res.in

V. P. Uniyal, Scientist, Wildlife Institute of India, Post Box # 18, Chandrabani, Dehradun 248001, Uttarakhand, India. Email uniyalvp@wii.gov.in

Zeenat Haq, Department of Botany, Pir Mehr Ali Shah Arid Agriculture University, Rawalpindi, Pakistan

Zubaida Yousaf, Ph.D. Scholar, Department of Botany, Lahore College for Women University Lahore, Pakistan. Email zubaida_yousaf@yahoo.com

Contents

Foreword *v*
Preface *vii*
List of Contributors *xiii*

Section I: Assessment of Biodiversity

1. **Fungal Biodiversity in Forests: Current Knowledge and Future Challenges** **3**
   Introduction 3
   Materials and Methodology 4
   Results and Discussion 4
   Diversity of Fungi in India 5
   Fungi Communities and Ecology 8
   Fungal Biodiversity Conservation 8
   Conclusion *9*
   References *10*
2. **Spiders (Araneae) in Bhyundar Valley of the Nanda Devi Biosphere Reserve** **13**
   Introduction 13
   Materials and Methodology 15
   Results 17
   Discussion 21
   Acknowledgements 22
   References 22

3. **Status of the Forest and Wildlife in the Binsar Wildlife Sanctuary** **25**
   Introduction 25
   Methodology 26
   Results 28
   Discussion 30
   References *34*
4. **Population and Management of Medicinal Plants in Chhattisgarh** **37**
   Introduction 37
   Study Area 38
   Methodology 39
   Results and Discussion 40
   Acknowledgements *48*
   References *48*
5. **Application of Remote Sensing and GIS in Mapping the Plant Resources of Odisha** **51**
   Introduction 51
   Study Area 55
   Materials and Methodology 57
   Results and Discussion 59
   Acknowledgements *67*
   References *68*
6. **Chemical Impact of *Parthenium hysterophorus* Linn. on Some Economically Important Vegetables** **71**
   Introduction 71
   Materials and Methodology 72
   Statistical Analysis 73
   Results 73
   Discussion 85
   Conclusion *88*
   References *88*

7. **Genetic Factors Contributing to Plants Vulnerability and Extinction** **91**

Introduction 91

Genetic Characters and Extinction 92

Comparison of Rare and Widespread Species 97

Implications for Conservation 99

Conclusion *101*

References *101*

8. **Use of Geospatial Techniques in Monitoring and Management of Biodiversity of Urban Ecosystems** **113**

Introduction 113

Monitoring Urban Biodiversity 115

Urban Biodiversity and Sustainability 119

RS–GIS Applications in Urban Biodiversity Management 130

Future Research 133

Conclusion *137*

Acknowledgements *138*

References *138*

Section II: Community, Livelihood, and Management of Biodiversity

9. **From Cabbage to Conservation: Case of Community Forest Conservation by Hmong Tribe in Northern Thailand** **149**

Introduction 149

Study Area 150

Methodology 151

Results 152

Discussion 160

Conclusion *161*

Acknowledgements *161*

References *162*

10. Increasing Livelihood Security through Kuth Cultivation in Tons Watershed, Garhwal Himalaya **165**
Introduction 165
Uttarakhand: A State Of Medicinal and Aromatic Plants 167
Collection practices of Medicinal and Aromatic Plants 169
Cultivation of Medicinal and Aromatic Plants 170
Role of SHER in farming of Kuth 174
References *180*

11. Case Study of Traditional Agroecosystems in Patharkot Village, Uttarakhand **183**
Introduction 183
Study Area 184
Methodology 185
Structure of the System 185
Conclusion *190*
References *193*

12. Conservation and Livelihood Links of Joint Management of Protected Area in Phutoei National Park, Thailand **195**
Introduction 195
Study Area 196
Methodology 197
Results 198
Discussion 206
Conclusion *208*
References *208*

13. Sustainable Utilization and Management of Timur by Communities in Uttarakhand **211**
Introduction 211
Materials 212
Study Area 213

Methodology 214
Results 215
Discussion 222
Conclusion *224*
Acknowledgement *224*
References *225*

14. Ethno-taxonomical Study of Medicinal Plants in the Nanda Devi Biosphere Reserve **227**
Introduction 227
Methodology 229
Results and Discussion 230
Conclusion *240*
Acknowledgements *241*
References *241*

15. Indigenous Knowledge, Use of NTFPs, and Biodiversity Conservation in Uttarakhand **245**
Introduction 245
Study Area 246
Methodology 246
Results and Discussion 247
Conclusion *253*
References *254*

Section III: Climate Change and Biodiversity Conservation

16. Climate Change and Challenges of Biodiversity Conservation **259**
Introduction 259
Effects of Climate Change on Biodiversity 260
Climate Change Mitigation Strategies and CDM Projects 265
References *267*

17. Role of Forests and Associated Traditional Knowledge of Local Communities in Climate Change Adaptation **271**

Introduction 271

Potential Impacts of Climate Change on Forests and Forest-dependent Communities 272

Sustainable Forest Management and Climate Change Adaptation 273

Indigenous Knowledge on Coping Strategies 275

Discussion and Conclusion 278

References *279*

18. Adaptation and Mitigation Strategies of Climate Change for Sustainable Agriculture **283**

Introduction 283

Brief Overview of Climate Change in India 284

Linkages Between Agriculture and Climate Change 285

Impacts of Climate Change on Agriculture 286

Climate Change Concern in Horticulture 286

Animal Husbandry and Climate Change 288

Climate Change Concern in Fisheries 288

Adaptation to and Mitigation of Climate Change in the Agriculture Sector 289

Suggestive Adaptation and Mitigation Strategies 290

Conclusion *295*

Acknowledgements *296*

References *296*

19. Climate Change Mitigation: Potential of Growing Tree Species of Medicinal Value **299**

Introduction 299

Study Area 301

Methodology 302

Results 306

Net Present Value 309

Discussion and Conclusion 309
Acknowledgements *313*
References *314*

Index *317*
About the Editors *329*

Section I

Assessment of Biodiversity

CHAPTER 1

Fungal Biodiversity in Forests: Current Knowledge and Future Challenges

Kavita Sharma and R N Pati

INTRODUCTION

Fungal biodiversity has become an integral part of human welfare. Forests are biologically diverse systems, representing some of the richest biological areas on Earth. They offer a variety of habitats for plants, animals, and microorganisms. Forests form the source for timber, pulpwood, firewood, fodder, meat, cash crops, fish, and medicinal plants and provide livelihoods for hundreds of millions of people worldwide. But only a fraction of the known species has been examined for potential medicinal, agricultural, or industrial value. Microbes constitute a major part of this biodiversity and are also a potential source of useful bioactive compounds. Microbial biodiversity provides the foundation for biotechnology, bioremediation, and genetic manipulation of organisms for new commercial products and processes. Thus, retention and conservation of microorganisms for future applications is of great importance for global economy.

This study was done in the Dongargarh region of Chhattisgarh. The state has a rich biodiversity and is covered by a vast forest area that comprises sal forests, teak forests, and other miscellaneous forests. The microbes inhabiting these forests are capable of damaging leaves, stem roots, stored wood, flowers, and seeds. The decomposition of the forest flora is accelerated by microbial activity that ultimately releases nutrients and maintains the nutrient dynamics for the plant community. Many of the higher fungi, belonging to different taxonomic groups, are responsible for decaying wood, some of which spoil timber beyond the economic threshold level, and are thus categorized as harmful microbes. Some of the higher fungi are responsible for decomposing the woody plant material into humus and help maintain the carbon and nitrogen ratio in the soil. Fungal spores make up a major part of the air spora. They may originate in the air from various sources such as plant, water, and soil and may settle down on a suitable host for growth and development. Over the last few years, significant progress has been observed in the study of airborne fungi due to the medical and phytopathological consequences associated with fungal spores (Picco and Rodolfi 2000). The aim of the present investigation is to understand the diversity of aeromycoflora in the forest area of Dongargarh.

MATERIALS AND METHODOLOGY

The survey was conducted for a year, from June 2009 to May 2010, during which fungi were collected from the Dongargarh forest floor. The fungi were isolated using petri plate method containing potato dextrose agar (PDA) media (Grover, Kumar, Kumar, *et al.* 2007) and incubated at $26 \pm 1°C$ for 6 to 8 days. At the end of the incubation period, the fungal colonies were counted. The fungi were identified morphologically and microscopically by National Centre of Fungal Taxonomy, Delhi and the percentage contribution of each fungal species was calculated.

RESULTS AND DISCUSSION

Fungal Diversity of Donagargarh

The results of the aeromycoflora survey showed that the forest atmosphere is never free of fungal spores. During the present investigation, a total of 177 colonies belonging to 21 fungal species were isolated from the Dongargarh forest area. The maximum percentage contribution was observed for *Aspergillus niger* (11.29%), followed by *Curvularia lunata* (10.16%), *Alternaria* sp. (8.47%), *Aspergillus fumigatus* (6.77%), and the lowest for *Trichoderma viridi* (1.69%). The fungal spores recorded were representatives of three major groups: Zygomycotina, Anamorphic fungi, and *Mycelia sterilia*. The predominant mycoflora in the forest belonged to the Anamorphic group of fungi representing 18 fungal species, followed by Zygomycotina (two fungal species) and one species of *M. sterilia* (Table 1). Figure 1 shows *C. lunata* and Figures 2(a) and 2(b) show *A. niger* and *Fusarium* as the fungi isolated from the Dongargarh forest.

The results of the investigation were compared with the work/findings of other researchers. Some *Aspergillus* species such as *A. niger* and *A. tamari* were isolated from the forest soil and cave ecosystems of Taiwan by Agoramoorthy and Hsu (2001). Arunachalam, Arunachalam, Pandey, *et al.* (1997) studied the dynamics of microbial population in subtropical humid forest in north-east India and observed that *Aspergillus* and *Penicillium* were the abundant genera in the sites. Das and Gupta Bhattacharya (2007) studied airborne culturable fungal flora of an agricultural farm in West Bengal and reported that *Cladosporium*, *Aspergillus*, and *Curvularia* were the predominant fungal species. A similar observation was obtained by Sharma and Tiwari (2008). Uddin (2004) reported in West Bengal that *Penicillium* and *Aspergillus* were the most dominant saprophytes.

DIVERSITY OF FUNGI IN INDIA

As there is only limited information available about the biodiversity of fungi in India, it is imperative to search for new forms of fungi and conserve them. There are three ways by which fungi can be conserved and are as follows.

1. Better understanding of fungal diversity and fungal ecology.
2. Improved awareness across society about the importance of fungi and better articulation about their conservation.
3. Documenting the traditional knowledge on the use of fungi and passing on the same to future generation.

Table 1 *Observations of different fungal species during the study period (June 2009–May 2010)*

Name of fungi	*Contribution (%)*
Zygomycotina	
Mucor sp.	5.41
Rhizopus sp.	4.51
Anamorphic fungi	
Alternaria alternata	8.47
Alternaria sp.	3.38
Aspergillus flavus	4.51
Aspergillus fumigatus	6.77
Aspergillus japonicus	3.38
Aspergillus luchensis	3.38
Aspergillus niger	11.29
Aspergillus tamari	2.25
Cladosporium cladosporioides	5.64
Cladosporium sphaerospermum	3.38
Curvularia lunata	10.16
Curvularia sp.	4.51
Fusarium sp.	3.38
Nigrospora oryzae	3.38
Penicillium chrysogenum	4.51
Penicillium citrinum	3.38
Phoma glomerata	3.38
Trichoderma viride	1.69
Mycelia sterilia	2.82

In India, many researchers have contributed in the study of airborne fungi. For example, Kulkarni and Tilak (1972) studied the microbial content of air inside and outside the caves at Aurangabad. Ponnusamy and Vittal (1979) made a preliminary study of the atmospheric fungal flora in Madras, while Ramalingam (1971) had done a similar one in Mysore. Prasad and Narayan (1982) studied the aeromycology of the atmosphere of Malleswaram market in Bengaluru. Babu and Singh (1983) reported on the airborne fungal flora of Delhi. Manoharacharya, Tulsi Reddy, and Vijaygopal (1988) conducted aeromycological studies in some localities of Andhra Pradesh. Jadhav and Tiwari (1994) observed aeromycoflora in Ravan village,

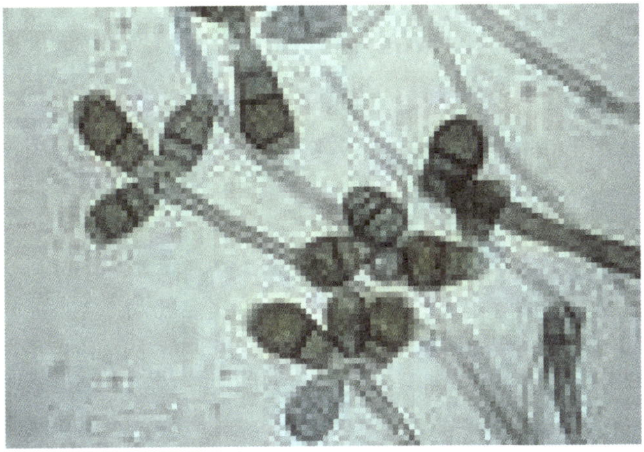

Figure 1 *Curvularia lunata*

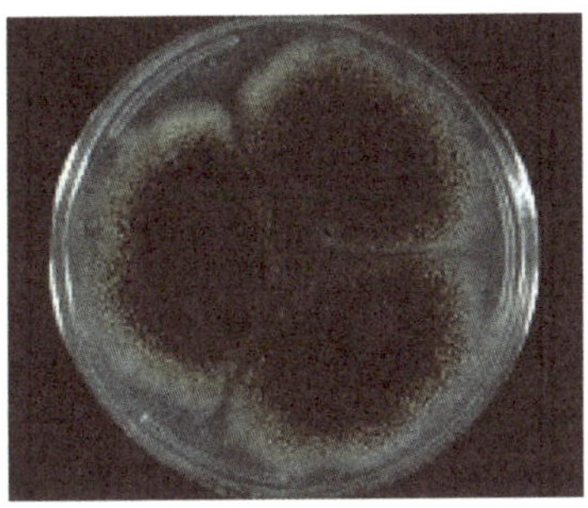

(a) *Aspergillus niger*

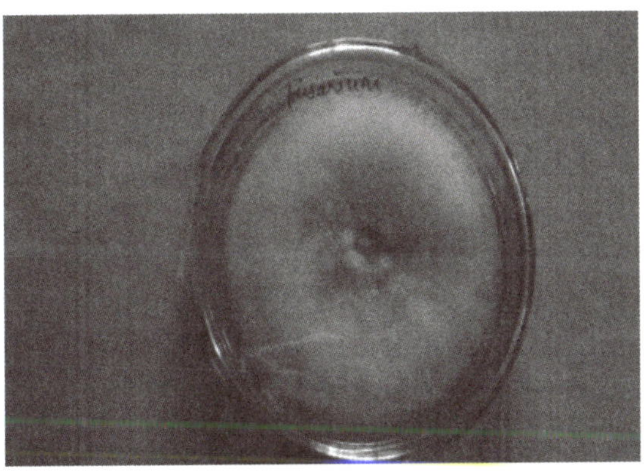

(b) *Fusarium*

Figure 2 Fungi isolated from Dongargarh forest

Chhattisgarh. Arora (1995) conducted an aerobiological survey of Churu district, Rajasthan. Sharma (2001) studied aeromycoflora in relation to leaf surface mycoflora in *Ocimum sanctum* L. Bhagyalaxmi and Saroja (2003) conducted preliminary studies on aeromycology in Hyderabad. Naveen and Puttaiah (2007) studied the aeromycoflora of Bhadrawati town, Karnataka.

FUNGI COMMUNITIES AND ECOLOGY

Fungi are known to play a vital role as decomposers, symbionts of plants and animals, and as parasites of plants in different ecosystems. Fungi interact with their hosts and with abiotic variables in the environment. They occur on rocks, in soil, in the sea and freshwater, in extreme habitats experiencing high and low temperatures, on dry substrates, and in concentrated nutrients. Members of Mucorales are considered ruderals since they survive in soil as long as nutrients are available, although they are not capable of degrading cellulose or lignin. Fungi like *Fusarium*, *Penicillium*, and *Trichoderma* are stress tolerant. The majority of fungi are mesophiles, with maximum growth between 25°C and 30°C (*Mucor* sp. *Penicillium chrysogenum*). Xerotolerant fungi can grow on dry material (*Aspergillus*, *Pencillium*).

FUNGAL BIODIVERSITY CONSERVATION

Geographic location, climatic conditions, microhabitat, substrate type, and distribution of fauna and flora are all important factors contributing to fungal distribution around the world. The challenges for fungal conservation are daunting. As nature's recyclers, fungi are like municipal refuse collectors employed to take away our rubbish. Therefore, the well-being of fungi is necessary for sustainable life on this planet. Scientists have known for over 100 years that, like animals and plants, fungi, too, are affected by the destructive activities of mankind. The impact of air pollution on lichen-forming fungi is one particularly well-documented example. Although there is still insufficient information about the conservation status of fungi, there is no reason to suppose that fungi are any less vulnerable, than other groups of organisms, to habitat loss and climate change.

All over the world, the threat to fungi is of concern since they are not only beautiful but also play a significant role in the human welfare. Moore, Nauta, Evans, *et al.* (2001) suggest the following three steps for fungal conservation.

1. Conservation of habitats.
2. In-situ conservation of non-mycological reserves/ecological niches.
3. Ex-situ conservation, especially for saprotrophic species growing in culture.

Fungi (for example, host-specific species known only on rare endemic plants) are often treated as part of the problem, and as a threat to the plant, rather than as being in need of protection themselves. In many countries, there is no explicit legal protection for fungi. Till date, mycologists have concentrated on gathering scientific evidence of population decline of fungi. But fungal conservation is more than just scientific evidence. It has attracted the attention of mycologists since the early 1980s, when reports started appearing on the prominent decrease of certain groups of macrofungi in some countries. Today there is a greater need for accurate information on the ranges of habitat types, substrates, and host specificity of species in different parts of Europe, preferably in a semi-quantitative form.

Fungi are omnipresent in the environment. Studies of fungal distribution and mapping are challenging tasks due to lack of sufficient taxonomic knowledge. Studies on fungal diversity in forests are very rare. Thus, there is a need to investigate and list the fungi from different soils and habitats in the forest ecosystem and to identify the major functional group of microorganisms responsible for the biological diversity of macroorganisms. Since there is no specific listing of the fungal diversity of Chhattisgarh, it is necessary to prepare a project on fungal diversity and to list out the significant microbes in different habitats.

CONCLUSION

Unlike prokaryotes, automated isolation techniques for fungi are not yet possible because of the extreme diversity in fungal mycelium, growth, and texture. The world of fungi provides a fascinating and almost endless source of biological diversity. The fungal diversity of any forest depends on a large number of factors concerning the soil, such as pH, organic contents, and moisture. Species diversity of fungi is also related to particular habitats and ecosystem. A total of 21 fungal species were observed in the Dongargarh forest during the present investigation. Some common aeroallergens including *Aspergillus* and *Penicillium* were found to dominate the mould spectrum. This is only a preliminary study, and is open to confirmation and refinement with more data from future studies.

REFERENCES

Agoramoorthy, G., and M. J. Hsu. 2001. Occurrence and diversity of thermophilous soil microfungi in forest and cave ecosystems of Taiwan. *Fungal Diversity* 7:27–33

Arora, A. 1995. Pollen morphology and aerobiological survey of Churu district of Rajasthan. [PhD thesis submitted to MDS University, Ajmer]

Arunachalam, A., K. Arunachalam, H. N. Pandey, and R. S. Tripathi. 1997. Dynamics of microbial population during the aggradation phase of a selectively logged subtropical humid forest in northeast India. *Tropical Ecology* 38:333–341

Babu, C. R., and A. B. Singh. 1983. Airborne fungal flora of Delhi. *The Indian Journal of Chest Disease and Allied Science* 25:31–35

Bhagyalakshmi, O., and P. V. Saroja. 2003. Preliminary studies on aeromycology of Hyderabad (Andhra Pradesh). *Indian Journal of Aerobiology* 16(1&2):59–60

Das, S., and S. Gupta Bhattacharya. 2007. Airborne culturable fungal flora of an agricultural farm in West Bengal and its relationship with meteorological factors. *Indian Journal of Aerobiology* 20(1&2):1–8

Grover, R., P. Kumar, S. Kumar, and K. P. Sharma. 2007. Response of fungal community in the unpolluted and polluted habitats. *Journal of Environment and Science Engineering* 49(2):93–98

Jadhav, S. K., and K. L. Tiwari. 1994. Aeromycoflora of Ravan village. *Indian Botanical Report* 1(1):33–36

Kulkarni, R. L., and S. T. Tilak. 1972. Microbial contents of air inside and outside the caves at Aurangabad. *Current Science* 23:850–851

Manoharacharya, C., V. R. Tulsi Reddy, and K. Vijaygopal. 1988. Aeromycological studies from some localities of Andhra Pradesh, India. *Biome* 3:25–30

Moore, D., M. M. Nauta, S. E. Evans, and M. Rotheroe. 2001. *Fungal Conservation: Issues and Solutions.* Cambridge, MA: Cambridge University Press

Naveen, D., and E. T. Puttaiah. 2007. Airmycoflora of Bhadravati town, Karnataka, India. *Environment and Ecology* 25(3):561–563

Picco, A. M., and M. Rodolfi. 2000. Airborne fungi as biocontaminants at two Milan underground stations. *International Biodeterioration and Biodegradation* 45:43–47

Ponnusamy, P. A., and B. P. R. Vittal. 1979. Preliminary study of the atmospheric fungal flora of Madras. *Kavaka* 7:78–82

Prasad, J., and M. C. Narayan. 1982. Aeromycology of the atmosphere of Malleswaram market, Bangalore. *ActaBotanica India* 10:196–200

Ramalingam, A. 1971. Air-spora of Mysore. *Proceedings of the Indian Academy of Science* B 74:227–240

Sharma, K. 2001. Studies of aeromycoflora in relation to leaf surface mycoflora of *Ocimum sanctum* Linn. plant. [PhD thesis submitted to Pundit Ravi Shankar Shukla University, Raipur, Chhattisgarh]

Sharma, K., and K. L. Tiwari. 2008. Leaf surface mycoflora of *Ocimum sanctum. Plant Archives* 8(2):667–669

Uddin, N. 2004. Airspora Rabi season in the studies over a rice field in West Bengal. *Aerobiologia* 20(2):127–134

CHAPTER 2

Spiders (Araneae) in Bhyundar Valley of the Nanda Devi Biosphere Reserve

Shazia Quasin and V P Uniyal

INTRODUCTION

Spiders are a special group of invertebrates that exploit a wide variety of niches in virtually all biomes on earth and play a vital role in sustaining the ecosystem. They are a diverse group of animals attaining seventh number in diversity (Benz and Nyffeler 1980). Spiders, which include 42055 described species under 3821 genera and 110 families globally, comprise a significant portion of terrestrial arthropod diversity (Platnick 2011). Since the early 1980s, a marked number of taxonomic and faunistic reports on arctic, boreal, and temperate spiders, both in the Palaearctic and in the Nearctic regions, have been published. More than 34000 spider species, of an estimated total of 170000 extant species, have been described (Coddington and Levi 1991). In the US and Canada alone (Roth 1993) approximately 3500 described species (and an additional estimated 350 undescribed species) exist.

In India, 1442 species of spiders belonging to 361 genera of 59 families have been reported so far. Of these, 21 genera are endemic to India while 13 are endemic to South Asia. Of the 1442 species, 1002 are endemic to the Indian mainland, 71 species are endemic to South Asia, 65 species are endemic to the Andaman and Nicobar Islands, and 1 species is endemic to Lakshadweep (Siliwal, Molur, and Biswas 2005). The Indian spiders checklist prepared by Siliwal, Molur, and Biswas (2005) includes nine subspecies of spiders, of which seven are endemic to the Indian mainland and two to the Andaman and Nicobar Islands. The families represented by the highest number of genera and species in India are *Salticidae* (62 genera and 181 species), followed by *Thomisidae* (37 genera and 154 species).

Spiders are important regulators of insect population (Lockley and Riechert 1984; Bishop and Riechert 1990; Wise 1993) and may prove to be useful indicators of overall species richness and the health of the biotic communities (Colwell, Erwin, Kremen, *et al.* 1993; Coddington and Colwell 1994; Norris 1999). Spider communities are sensitive to a wide range of environmental factors, including habitat structure (Uetz 1991), habitat type (Eyre and Rushton 1992; Rushton 1988), and exposure to wind, moisture, and temperature (Wise 1993).

Although they are a relatively difficult group, taxonomically they have the potential to be ecological indicators (Alderweireldt, Baert, Desender, *et al.* 1989) where suitable taxonomic skills are available and may be effective in assessing and monitoring the ecosystem. The fact that spiders are abundant in nature (Wise 1993), easy to collect, and found in varied habitats make them suitable for biodiversity and bio-monitoring studies (Churchill 1997).

In India, knowledge about diversity and distribution of spiders is sparse in comparison to other regions of the world. Little information is available about them in the Northern part of India, especially in the higher altitudinal zones of the Himalayan region. An urgent need exists to explore the spider fauna in the Himalayan region of the country. Considering their role in the ecosystem, the present study aims to describe the spider diversity in the Bhyundar Valley of the Nanda Devi Biosphere Reserve (NDBR). This study, thus, attempts to make an inventory of the spider species, their density, composition, and richness in different sites of the NDBR with respect to altitudinal gradient.

MATERIALS AND METHODOLOGY

The Bhyundar Valley (30°37' N, 79°33' E) is located in the Nanda Devi Biosphere Reserve in the northern part of the Western Himalayas and comprises parts of Chamoli district in Garhwal, Uttarakhand (Figure 1). The Bhyundar Valley leads into the Valley of Flowers National Park. The topography, climate, and soil of this area support diverse habitats, species, communities, and ecosystems. There is a high percentage of native and unique flora and fauna in this area (Zoological Survey of India 1997).

Sampling

Spider samples were collected from 10 random sampling plots 10 m by 10 m in size along the altitudinal gradient of 1800 m to 2300 m in the Bhyundar Valley (NDBR). In each of the plots, spider sampling was done using pitfall traps and by five other collection methods, such as sweep netting, ground hand collection, aerial hand collection, vegetation beating, and litter sampling. To collect ground-dwelling spiders, nine pitfall traps (cylindrical plastic bottles of 9 cm diameter and 10 cm depth) were arranged within a 10 m × 10 m sampling plot in three horizontal and three vertical rows, each at 5 m distance from the nearest neighbours, thus forming four smaller grids of 5 m × 5 m (Figure 2) within the sampling plot. The traps were filled with liquid preservative (69% water, 30% ethyl acetate, and 1% detergent). Other methods were applied to collect web building, ambush, and ground runner spiders. Sampling was carried out from March to May 2009. The collected specimens were transferred to 70% alcohol for further identification. Voucher specimens were deposited at the Wildlife Institute of India, Dehradun.

Statistical Analysis

The number of individuals belonging to a family collected from each sampling plot was used to estimate the density of each spider family recorded in the study area. The densities of different spider families were then compared for any significant difference using the one-way ANOVA in program *SPSS version 16.0*. The diversity of

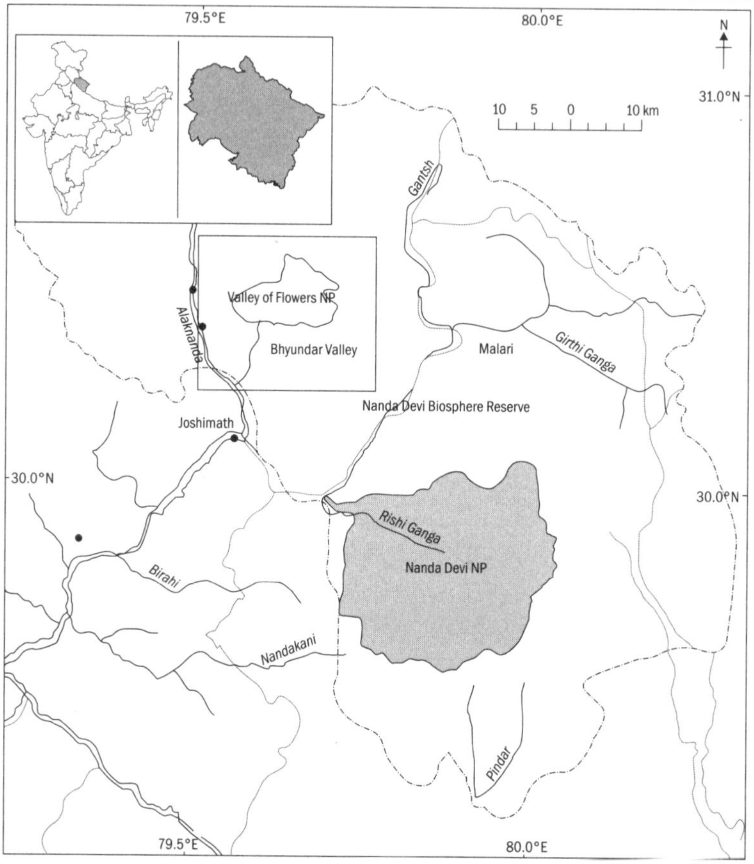

Figure 1 *Location of the study area: Bhyundar Valley*

spider families along the altitudinal gradient was analysed by the Shannon–Wiener index (H′), which is sensitive to changes in the abundance of rare species in a community (Magurran 1988) in program *ESTIMATE S version 8.0* (Colwell 2006). Based on a prior ecological knowledge, it was hypothesized that spider taxa richness would vary along altitude. Consequently, it was tested for normality of data through the Kolmorogov–Smirnov test and investigated for any relationship between family richness (response variable) and altitudinal gradient (predictor variable) using linear regression analysis in program *SPSS version 16.0.*

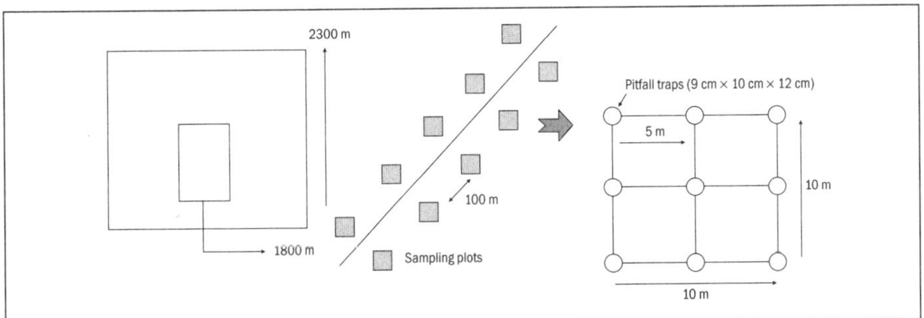

Figure 2 *Route map of the study area with the sampling plots*

RESULTS

Abundance, Richness, and Diversity

A total of 668 individuals (Table 1) belonging to 19 families, 34 genera, and 56 morpho-species/species were recorded. Out of the total 668 individuals, Araneidae was the numerically predominant family, comprising 22.9% of total spider abundance. It was followed by Lycosidae (13.9% of total spider abundance), Theridiidae (9.4%), Thomisidae (9.1%), Tetragnathidae (6.7%), Salticidae (6.1%), Gnaphosidae (5.68%), and Linyphiidae (5.6%). All the other families composed less than 5% of the overall abundance (Figure 3).

Composition of Spider Fauna

Families with the highest number of total species were the orb-web spiders Araneidae with 12 species (22% of all the species), followed by crab spiders Thomisidae (6 species; 11%), comb-footed spiders Theridiidae (5 species; 9%), jumping spiders Salticidae (4 species; 7%), and long-jawed spiders Tetragnathidae (4 species; 7%), while other families composed 24% of the total species richness with 13 species. In the lower altitudinal plots (1800–2100 m), 19 families of spiders were recorded from 10 plots, covering most of the families found in the study area. However, in the higher altitudinal plots (2101–2300 m), the number of families recorded from 10 plots decreased to seven, and mainly comprised Araneidae and Lycosidae (Figure 4).

Table 1 *Provisional checklist of spiders from Bhyundar Valley*

Family	*Genus/species*	*Guild*	*Method of collection*					
			PFT	*AHC*	*GHC*	*LS*	*VB*	*SN*
Agelenidae	*Agelena* sp.	WB	√	–	–	–	–	–
Araneidae	*Araneus* sp. 1	WB	–	√	–	–	√	√
	Araneus sp. 2	WB	–	√	–	–	√	√
	Araneus nympha	WB	–	√	–	–	–	√
	Araneus ellipticus	WB	–	√	–	–	–	√
	Araneus insutus	WB	–	√	–	–	–	√
	Cyclosa confraga	WB	–	√	–	–	–	√
	Cyclosa insigna	WB	–	√	–	–	–	√
	Cyrtophora moluccensis	WB	–	√	–	–	√	√
	Eriophora sp. 1	WB	–	√	–	–	–	√
	Neoscona muhkerjee	WB	–	√	–	–	–	√
	Neoscona theisi	WB	–	√	–	–	–	√
	Zygiella indica	WB	–	√	–	–	√	√
Clubionidae	*Clubiona* sp. 1	GW	√	–	√	–	–	–
	Clubiona sp. 2	GW	√	–	–	–	–	–
Gnaphosidae	*Gnaphosa* sp. 1	GW	√	–	√	√	–	–
	Gnaphosa sp. 2	GW	√	–	√	√	–	–
	Drassodes sp.	GW	√	–	–	–	–	–
Linyphiidae	*Linyphia* sp. 1	WB	–	√	–	–	–	√
	Linyphia sp. 2	WB	–	√	–	–	√	√
	Linyphia sp. 3	WB	–	√	–	–	√	√
Lycosidae	*Hippasa greenalliae*	WB	–	–	√	–	–	–
	Lycosa sp. 1	GW	√	–	√	√	–	–
	Lycosa sp. 2	GW	√	–	√	√	–	–
	Lycosa sp. 3	GW	√	–	√	–	–	–
	Pardosa sp. 1	GW	√	–	√	–	–	–
	Pardosa sp. 2	GW	√	–	√	√	–	–
Miturgidae	*Cheiracanthium* sp. 1	GW	–	–	√	–	–	–
Oxyopidae	*Oxyopes* sp. 1	PW	–	√	–	–	–	–
Philodromidae	*Philodromus* sp. 1	GW	–	–	√	–	–	–
	Philodromus sp. 2	GW	–	–	√	–	–	–
Pholcidae	*Crossopriza* sp. 1	WB	–	√	–	–	–	–
Pisauridae	*Pisaura* sp. 1	GW	√	–	√	–	–	–
	Pisaura sp. 2	GW	√	–	√	√	–	–

Contd...

Table 1 *Contd...*

Family	*Genus/species*	*Guild*	*Method of collection*					
			PFT	*AHC*	*GHC*	*LS*	*VB*	*SN*
Salticidae	*Plexippus* sp. 1	GW	√	–	√	–	–	–
	Myrmarachne sp. 1	GW	–	–	√	–	–	–
	Carrhotus sp. 1	GW	√	–	√	√	–	–
	Carrhotus sp. 2	GW	√	–	√	–	–	–
Selenopidae	*Selenops* sp. 1	GW	–	–	√	–	–	–
Sparassidae	*Heteropoda* sp. 1	GW	–	–	√	–	–	–
	Heteropoda sp. 2	GW	√	–	√	–	–	–
	Heteropoda venetoria	GW	–	–	√	–	–	–
Tetragnathidae	*Leucauge decorata*	WB	–	√	–	–	√	√
	Leucauge sp. 1	WB	–	√	–	–	√	√
	Leucauge sp. 2	WB	–	√	–	–	–	√
	Tetragnatha sp. 1	WB	–	√	–	–	√	√
Theridiidae	*Achaearanea* sp. 1	WB	–	√	–	–		√
	Achaearanea sp. 2	WB	–	√	–	–	√	√
	Theridion sp. 1	WB	–	√	–	–	√	√
	Theridion sp. 2	WB	–	√	–	–	√	√
	Theridion sp. 3	WB	–	–	–	–	√	√
Thomisidae	*Misumena* sp. 1	PW	–	√	–	–	–	√
	Misumena sp. 2	PW	–	√	–	–	–	√
	Runcinia sp. 1	PW	–	√	√	–	–	√
	Xysticus sp. 1	GW	–	√	√	√	–	–
	Diaea sp. 1	PW	–	√	√		√	–
	Diaea decorata	PW	–	√	–	–	–	√
Uloboridae	*Uloborus* sp. 1	WB	–	√	–	–	–	–
Zodariidae	*Lutica* sp. 1	GW	–	–	√	–	–	–

Guild structures: WB - web builders; PW - plant wanderers; GW - ground wanderers

Collection methods: PFT - pitfall traps; GHC - ground hand collection; AHC - aerial hand collection; VB - vegetation beating; LS - litter sampling; SN - sweep netting

Family richness showed a linear declining relationship with increasing altitudinal gradient (R^2 = −0.9089, P < 0.001; Figure 5).

Pearson's correlation indicated a significant negative correlation between family richness and altitude (R = −0.95). The Shannon-Weiner index estimated family diversity at 2.31 ± 0.01. The density of the families Araneidae and Lycosidae

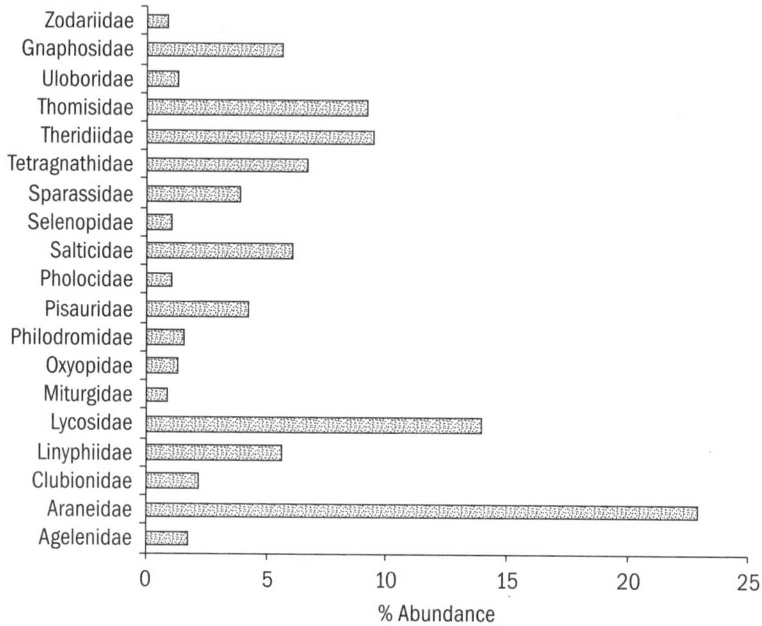

Figure 3 *Relative abundance of the spider families*

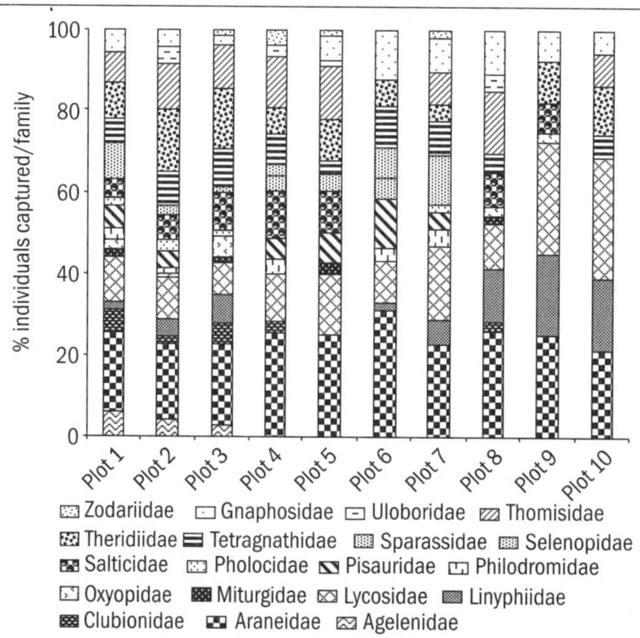

Figure 4 *Family composition of spiders in Bhyundar Valley*

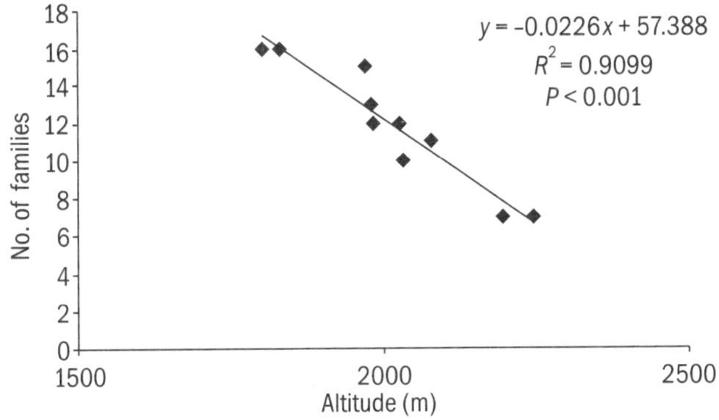

Figure 5 *Relationship between spider family richness and altitudinal gradient in Bhyundar Valley*

Note *Family richness declined linearly with altitude following the equation:* $y = -0.0226x + 57.366$, $R^2 = 0.9089$, $P < 0.001$

(0.15 ± 0.013 and 0.093 ± 0.009) was found to be significantly higher than that of the other families (F-test = 20.74, $P < 0.001$; Figure 6).

DISCUSSION

During the study, 19 families of spiders, which represent 31% of the total families reported from India (Sebastian and Peter 2009),

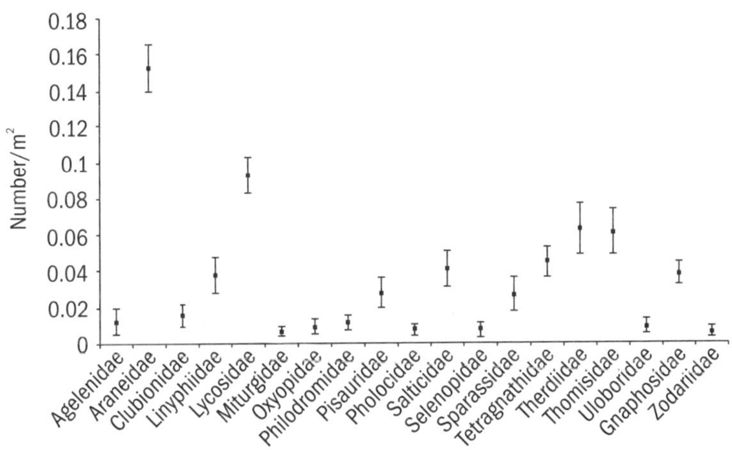

Figure 6 *Density (individuals/m²) of spider families in Bhyundar Valley*

were recorded. The lower altitudinal areas exhibiting high structural complexities, with a greater availability of niches, recorded highest spider diversity. The density of the families *Araneidae* and *Lycosidae* was found to be significantly more than the rest of the families, which was thought to be due to greater adaptability and resource availability for these families in the study area. However, some families such as *Lycosidae*, which are more tolerant and can overcome harsh conditions, were also collected from relatively higher elevations.

The results showed that the number of families recorded from the sampling plots linearly decreased with increasing altitude and also found that the family richness showed a consistent negative correlation with altitude. However, as only a small altitudinal range was tested in this study, this result may not be true for a larger altitudinal range. In order to test the effect of altitude on spiders, a larger altitudinal range must be taken into consideration. As spiders are sensitive to small changes in the environment, especially vegetation, topography, and climate, these parameters might show interesting effects on the spider diversity. It would be interesting to study the effect of these parameters on the spider fauna.

ACKNOWLEDGEMENTS

The authors thank Mr P R Sinha, Director, and Dr V B Mathur, Dean, Wildlife Institute of India for their support and guidance during this study. The authors would like to express their sincere gratitude to the Uttarakhand Forest Department for providing logistics and permission to carry out the study. The authors also thank Dr Sebastian Peter, Department of Zoology, Sacred Heart College, Cochin, for helping in the identification of spiders and for useful discussion. Financial support was provided by the Department of Science and Technology (SERC), New Delhi.

REFERENCES

Alderweireldt, M., L. Baert, K. Desender, and J. P. Maelfait. 1989. Lycosid spiders of the Belgian coast. *Bulletin Annales Societé Royal Belge Entomologie* 125:327–332

Benz, G., and M. Nyffeler. 1980. The role of spiders as insect predators in cereal fields near Zurich (Switzerland). In *Proceedings of the 8th International Congress of Arachnology*, pp. 127–131

Bishop L., and S. Riechert. 1990. Prey control by assemblage of generalist predators: spiders in garden test systems. *Ecology* 71:1441–1450

Churchill, T. B. 1997. Spiders as ecological indicators: an overview for Australia. *Memoirs of the Museum of Victoria* 56:331–337

Coddington, J. A., and H. W. Levi. 1991. Systematics and evolution of spiders (Araneae). *Annual Review of Ecology and Systematics* 22:565–592

Coddington, J. A., and R. K. Colwell. 1994. Estimating terrestrial biodiversity through extrapolation. *Philosophical Transactions of the Royal Society* (Series B) 345:101–118

Colwell, R. K. 2006. Estimate S: statistical estimation of species richness and shared species from samples. Version 8. Details available at <www.purl.oclc.org/estimate>

Colwell, R. K., T. L. Erwin, C. Kremen, D. D. Murphy, R. F. Noss, and M. A. Sanjayan. 1993. Terrestrial arthropod assemblages: their use in conservation planning. *Conservation Biology* 7(4): 796–808

Eyre, M. D., and S. P. Rushton. 1992. Grassland spider habitats in north-east England. *Journal of Biogeography* 19:99–108

Hänggi, A., and R. Maurer. 1991. Katalog der Schweizerischenspinnen. *Documenta Faunistica Helvetiae* 12:2–33

Lockley, T. C., and S. E. Riechert. 1984. Spiders as biological control agents. *Annual Review of Entomology* 29:299–320

Magurran, A. E. 1988b. *Ecological Diversity and its Measurement.* London: Croom Helm Ltd

Norris, K. C. 1999. Quantifying change through time in spider assemblages: sampling methods, indices, and sources of error. *Journal of Insect Conservation* 3:309–325

Platnick, N. I. 2011. *The World Spider Catalogue, Version 11.0.* New York: The American Museum of Natural History

Roth, V. D. 1993. *Spider Genera of North America*, 3rd edn. Florida: American Arachnological Society

Rushton, S. P. 1988. The effects of scrub management regimes on the spider fauna of chalk grassland at Castor Hanglands, National Nature Reserve, Cambridgeshire, UK. *Biological Conservation* 46:169–182

Sebastian, P. A., and K. V. Peter. 2009. *Spiders of India*, 1st edn. Hyderabad: Universities Press

Siliwal, M., S. Molur, and B. K. Biswas. 2005. Indian spiders (Arachnida: Araneae): updated checklist 2005. *Zoos' Print Journal* 20(10):1999–2049

Uetz, G. W. 1991. Habitat structure and spider foraging. In *Habitat Structure: The Physical Arrangement of Objects in Space*, edited by S. S. Bell, E. D. McCoy, and H. R. Mushinsky, pp. 325–348. London: Chapman and Hall

Wise, D. H. 1993. *Spiders in Ecological Webs.* Cambridge: Cambridge University Press

Zoological Survey of India (ed). 1997. *Fauna of Nanda Devi Biosphere Reserve: A World Heritage Site.* Dehradun: Zoological Survey of India

CHAPTER 3

Status of the Forest and Wildlife in the Binsar Wildlife Sanctuary

Chandra Prakash Kala and Bhagwan Singh Majila

INTRODUCTION

The Binsar Wildlife Sanctuary (BWS) lies in the Almora district of the Uttarakhand in the Indian Himalayas between 29°30' N to 29°43' N and 79°41' E to 79°47' E. The total geographical area of the sanctuary is 45.59 km^2. Being hilly, the sanctuary terrain is characterized by deep ravines, crevices, and elevated ridges. Geologically, BWS falls under the Inner Lesser Himalayan zone, which mainly comprises unfossiliferous sedimentary rocks of various ages, that is, from Precambrian to Paleozoic (Valdia 1976). The rocks are predominantly made up of the Korl group of Paleozoic age, which are composed of granite, granodiorite, graphite, schist, shale-quartzite, and quartzite (Gansser 1964). On an average, the soil is very rich in organic matter that contains a high proportion of plant nutrients (Majila 1992).

The monthly mean maximum temperature varies from 14.30°C (January) to 31.73°C (June) and the mean minimum temperature fluctuates from 0.20°C (January) to 12.30°C (August). During the coldest months, January and February, the study area received heavy dew at night. Snowfall usually starts in the middle of December and continues till the end of February. The monsoon generally begins after mid-June. Sometimes, however, it may start as early as late May or as late as mid-July and last till the end of September. The heaviest rainfall usually occurs in July and August (258.7–338.5 mm) and the annual rainfall is about 1041.8 mm.

The forested area begins from an elevation of 1600 m to the top of hills, about 2400 m. The remaining areas at lower altitudes are used for agriculture and livestock grazing. Crop farming and animal husbandry are the main economic activities of the local people living in and around the sanctuary area. Livestock provides the basis for livelihoods and is used for ploughing and composting crop fields. It also produces a variety of products such as milk, wool, and meat, which are important for socio-economic and cultural advancement.

The forested hilltops and slopes in the sanctuary are covered with Chir Pine (*Pinus roxburghii*), Banj Oak (*Quercus leucotrichophora*), and Rhododendron (*Rhododendron arboreum*) as pure stands or as a mixed forest. Pure pine forests are present between 1600 m and 1900 m, while mixed forests of pine and oak are distributed between 1900 m and 2100 m altitudes. Pure oak and mixed-oak forests (*Q. leucotrichophora* and *Quercus floribunda*) are present between 2100 m and 2400 m altitudinal range in the sanctuary (Majila and Kala 2010).

Relative humidity varies along aspects and types of vegetation. The average relative humidity varies on a monthly basis from 36% to 76% in the Chir Pine forest during April and August, respectively. In the oak forest, the relative humidity varies between 49% in April and 87% in August (Majila 1992; Majila, Joshi, and Kala 2005). The diverse climatic, geographical, and topographical conditions have shaped the diverse forest vegetation and wildlife species in BWS.

METHODOLOGY

Woody Vegetation

To study the vegetation, the Binsar Wildlife Sanctuary was stratified into three altitudinal ranges: 1600–1900 m, 1900–2100 m, and 2100–2400 m, and four geographical aspects: east, west, north, and south. All the four aspects at each altitudinal range were sampled to study the tree and shrub structures and their species diversity. Quadrats were laid in each aspect and altitudinal zone: 10 m × 10 m for trees and 5 m × 5 m for shrub species samplings. Trees occurring in quadrats were measured at 1.37 m from the ground level for their circumference at breast height (CBH). For the study of regeneration dynamics, a 5 m × 5 m permanent plot was marked on each aspect at all three stratified altitudes and the number of seedlings and saplings in it were enumerated. Frequent visits were made to these marked plots to observe the performance of young trees. Another sampling was done after a year. The difference in the two samplings was used to calculate the rate of mortality and natality. A transition matrix was prepared in terms of seedlings and saplings, as described by Horn (1975).

The collected data was quantitatively analysed for tree species diversity in different forest sites with the help of the Shannon–Wiener diversity index (H') (Shannon and Weaver 1949).

$$H' = -\Sigma(Ni/N)\log_2(Ni/N)$$

where Ni is the total number of individuals of a species and N is the total number of individuals of all species in a forest site.

Concentration of dominance (Cd) was calculated by Simpson's index (Simpson 1949).

$$Cd = \Sigma(Ni/N)^2$$

Wildlife

The secondary data on wildlife species available with the Wildlife Department of BWS were collected. Each year the Wildlife Department conducts wildlife surveys on various species. The records on livestock and humans in the sanctuary area were also collected from the department. Villagers were interviewed about the presence of wildlife species within the sanctuary. In order to reduce bias, the interview was held in a casual manner where notes were not taken. The wildlife guards and people in the neighbouring villages were also cross examined regarding animal–human conflicts within the sanctuary.

RESULTS

Status of Wildlife

The common leopard is a major wild predator in the sanctuary. Other predators include the jungle cat (*Felis chaus*) and the dhole (*Canis alpinus*). The major ungulate species include the goral (*Nemorhaedus goral*), the barking deer (*Muntiacus muntijak*), the serow (*Capricornis sumatraensis*), and the wild boar (*Sus scrofa*). The sanctuary also harbours diverse bird fauna, including the Black Francolin (*Francolinus francolinus*), Koklass Pheasant (*Pucrasia macrolopha*), Kalij pheasant (*Lophura leucomelanos*), Hill Partridge (*Arborophila torqueola*), Great Barbet (*Megalaima virens*), Hawk Eagle (*Spizeatus nipalensis*), Himalayan Griffon (*Gyps himalayensis*), Lammergeier (*Gypaetus barbatus*), and the yellow-billed magpie (*Cissa flavirostris*) (Khan, Sultana, and Hussain 2000).

Cattle lifting by common leopard is one of the main reasons for man-animal conflict in and around the sanctuary. Although it is difficult to estimate the number of cattle-lifting cases in areas where more than one species of predator, particularly large cats, operate. However, in the case of BWS, it was possible to conclude that the predator involved in cattle lifting as well as human attacks was the common leopard because of the dominance of this cat in the area. During 1996–2003, nine cases of injuries and one case of human killing due to wildlife attack were recorded by the Wildlife Department. No records are registered by the department prior to 1996. Most of such cases were due to leopard attacks and were reported either from agriculture fields adjoining the forest areas or from the forest areas. In the absence of wild prey species, leopards even tend to become man-eaters. The entire hilly region of Uttarakhand is historically known for incidences of man eating by leopards and the situation persists (Kala 1999). The killing and mauling of humans are major causes of conflict in the study area.

Distribution and Structure of Woody Species

There are marked changes in the distribution of plant species across the altitudinal gradient. At lower altitudes (1600–1900 m), *P. roxburghii* dominates the forest vegetation, except in the western aspect where *Q. leucotrichophora* is relatively higher in number. Of the shrub species, *Myrcine africana* is the most dominant shrub at lower altitudes, followed by *Berberis asiatica* and *Rhus parviflora*. The eastern and southern aspects at middle altitudes (1900–2100 m) are dominated by *P. roxburghii* whereas western and northern aspects are dominated by *Q. leucotrichophora*. *M. africana* dominates the shrub layer at the middle altitude, along with *Rubus ellipticus*, *Hypericum patulum*, and *B. asiatica*. The higher altitudinal range (2100–2400 m) is dominated by *Q. leucotrichophora* and *Q. floribunda*. Except in the western aspect where *Q. floribunda* dominates the tree cover, and the rest of the aspects at higher altitudes in the BWS are dominated by *Q. leucotrichophora*. The *M. africana* shrub layer dominates across the altitudinal range. *Berberis aristata*, *Arundinaria falcata*, *Daphne papyracea*, and *H. patulum* are the other dominant shrub species in the higher altitudes (Majila and Kala 2010).

Species Diversity and Dominance

Based on the Shannon–Weiner Index, the highest diversity of tree species ($H' = 3.38$) was noted at middle altitude on the western aspect, followed by higher altitude ($H' = 2.82$) and lower altitude ($H' = 2.78$) on the same aspect. The lowest species diversity is calculated in the lower altitude on the southern aspect ($H' = 1.01$). *Cd* for the tree species is highest at lower altitudes on the southern (0.64) and eastern (0.40) aspects, and lowest at the same altitude on the western aspect. The highest diversity of shrub species was calculated for higher altitudes on the northern aspect ($H' = 2.55$), followed by lower altitudes on the southern aspect ($H' = 2.42$). The middle altitude possesses less species diversity on all aspects in comparison to lower and higher altitudes, and ranges between $H' = 1.22$ and $H' = 1.50$. *Cd* for the shrub layer is highest at middle altitude on the southern (0.58), eastern (0.56), and western (0.55) aspects, and lowest at lower altitudes on the southern aspect (Table 1).

Table 1 *Species diversity (H′) and concentration of dominance (Cd) for tree and shrub layer on various aspects across the altitudinal gradient in the Binsar Wildlife Sanctuary*

Altitude	Aspect	Tree layer		Shrub layer	
		Species diversity (H′)	Concentration of dominance (Cd)	Species diversity (H′)	Concentration of dominance (Cd)
Lower	East	1.802	0.408	2.30	0.23
	West	2.784	0.182	2.31	0.28
	North	2.590	0.220	2.32	0.25
	South	1.013	0.644	2.42	0.22
Middle	East	2.406	0.245	1.43	0.57
	West	3.380	0.389	1.50	0.56
	North	2.134	0.274	1.22	0.52
	South	2.007	0.373	1.42	0.59
Higher	East	2.493	0.195	2.39	0.30
	West	2.821	0.221	2.31	0.28
	North	2.729	0.219	2.55	0.23
	South	2.571	0.231	1.55	0.55

Natality and Mortality of Seedlings and Saplings

The natality and mortality rates of seedlings and saplings of three dominant tree species (*P. roxburghii*, *Q. leucotrichophora*, and *Q. floribunda*, at lower, middle, and higher altitudes respectively) described in Table 2 show some differences across the aspects. The highest seedlings and saplings natality of *P. roxburghii* is estimated on the southern aspect at lower altitude, whereas the mortality of its seedlings is highest on the west and the mortality of its saplings is highest on the north. *Q. leucotrichophora* has the highest seedlings natality on the western aspect at middle altitude and lowest seedlings natality on all aspects at lower altitude. *Q. floribunda* has higher seedlings and saplings mortality than *P. roxburghii* and *Q. leucotrichophora*. Proportionately, the mortality rate of seedlings and saplings of all the three species are higher than their natality rates.

DISCUSSION

In the Binsar Wildlife Sanctuary, the altitude plays an important role in the distribution and composition of woody species. Changes

Table 2 *Percentage of natality (N) and mortality (M) of seedlings and saplings of tree dominant tree species on various aspects across the altitudinal gradient in the Binsar Wildlife Sanctuary*

Altitude	*Aspect*	*Pinus roxburghii*				*Quercus leucotrichophora*				*Quercus floribunda*			
		Seedling		*Sapling*		*Seedling*		*Sapling*		*Seedling*		*Sapling*	
		N	*M*	*N*	*M*	*N*	*M*	*N*	*M*	*N*	*M*	*N*	*M*
Lower altitude	East	37.32	62.68	52.45	47.55	6.13	93.87	45.23	45.23	–	–	–	–
	West	23.17	76.83	47.82	52.18	9.45	90.50	21.23	78.70	–	–	–	–
	North	32.38	67.62	49.53	50.47	6.78	93.90	12.38	87.62	–	–	–	–
	South	47.82	52.18	59.57	40.43	6.10	93.90	9.68	90.32	–	–	–	–
Middle altitude	East	30.56	69.44	54.15	45.85	13.16	86.84	38.90	61.10	–	–	–	–
	West	13.55	86.45	39.58	60.42	18.78	81.22	46.29	53.71	–	–	–	–
	North	28.39	71.61	37.62	62.38	13.86	86.14	44.78	55.22	–	–	–	–
	South	33.37	66.63	53.65	46.35	12.27	87.73	33.39	66.61	–	–	–	–
Higher altitude	East	–	–	–	–	14.68	85.32	45.25	54.75	13.78	86.22	45.45	54.55
	West	–	–	–	–	16.26	83.74	61.30	38.70	15.68	84.32	62.49	37.51
	North	–	–	–	–	16.26	83.04	47.65	52.35	15.26	84.74	44.42	55.58
	South	–	–	–	–	15.17	83.83	50.10	49.90	14.58	85.42	50.52	47.48

in tree species composition across the various forest types reflect the environmental changes along the altitudinal gradient, which is also responsible for forming different vegetation zones (Kala and Uniyal 1999). Studies conducted for other mountains of the world demonstrate a strong connection of changes in community associations to altitude (Lovett 1996). Due to higher moisture content, oak forests are rich and diverse compared to Chir Pine forests in the BWS. The middle altitude, being a transition zone (ecotone) between lower and upper altitudes, obtains high species diversity. Both the tree and the shrub species richness in the sanctuary decreases with altitude. The tree and shrub species diversity in the sanctuary is analogous to other areas of the Uttarakhand Himalayas such as Nainital oak forest ($H' = 1.04$; Uprety 1982) and Kumaon Himalayas ($H' = 2.82$; Singh 1989). The species diversity of the sanctuary is also comparable to the temperate forests ($H' = 2.00 - 3.00$) studied by Monk (1967), and Risser and Rice (1971). The high-species diversity of any given area is also subjected to a high degree of vertical and horizontal micro environmental heterogeneity or high niche differentiation (Kappelle, Uffelen, and Cleef 1995).

The diversity dominance relationship in BWS indicates that species diversity increased with decrease in *Cd* and vice versa. Thus, an inverse relationship was noted between diversity and dominance for shrub and tree layers. On the basis of density, the correlation coefficient between species diversity and dominance was $r = 0.94$ for tree species and $r = 0.97$ for shrub species. Based on various environmental characteristics, Singh and Singh (1987) have suggested that environmental conditions in the Uttarakhand Himalayas are favourable up to about 2500 m for tree growth. The regeneration of *P. roxburghii* was best at lower altitudes on the southern and eastern aspects, which shows that this species mostly regenerates on warm, dry slope. Similar observations were made for pine forests by Horn (1975). However, Rao (1984) pointed out that *P. roxburghii* could regenerate well in all kinds of habitats. In general, the regeneration potential in most tree species declines with increase in altitude. *Q. leucotrichophora* had a good number of seedlings and saplings across the altitudes, but its highest seedling and sapling concentration was in the middle altitude. This trend indicates that the saplings and seedlings also represent the dominant species in each altitudinal range.

Besides many biotic and abiotic factors, species composition, species diversity, and regeneration dynamics reflect the historical patterns of land use. The existence of forest communities across the altitudes and aspects has certain links with historical anthropogenic pressures as well as the current physical conditions. The low-species diversity in certain areas may also be due to early successional stages, low rainfall, and a locally fluctuating climate favouring taxa with broad tolerance or species-specific biological adaptations (Saxena and Singh 1982; Youngblood, Max, and Coe 2004). The low-species diversity and richness at lower altitudes may also be due to past disturbances in the form of severe human impact as villagers collected fuel, fodder, and timber from these areas before the area was marked as a wildlife sanctuary. Some workers have reported the absence or poor availability of a shrub layer in the Chir Pine forest (Ralhan, Saxena, and Singh 1982; Singh 1989; Kala 2004). Yet, in BWS, the pine forest supports a good growth of shrub species, which may be due to the present protection network.

Various land-use practices occur in BWS, including agriculture, livestock grazing, and forestry. These practices have negatively impacted wild ungulate populations. Since wild ungulate prey occurs in low densities in the sanctuary, wild carnivores resort to lifting livestock and sometimes prey on humans, as reported in the case of leopards. This is the root of severe conflicts between conservation agencies and local communities in the sanctuary area. To mitigate such conflicts, the Wildlife Department in the BWS has employed several strategies such as cash compensation, indirect compensation through integrated conservation and development programs, and selective sustainable extraction of resources. Despite all this, the involvement of the local people is mainly less recognized while framing policies for natural resources conservation and management. Improving the local people's understanding and keeping them involved in management decisions may cultivate a positive attitude towards conserving wildlife and forests.

REFERENCES

Gansser, A. 1964. *Geology of the Himalayas*. London: John Wiley and Sons

Horn, H. S. 1975. Markovian properties of forest succession. In *Ecology and Evolution of Communities*, edited by M. L. Cody and J. M. Dimond, pp.196–211. Cambridge: Harvard University Press

Kala, C. P. 1999. Man-eaters of Garhwal. *WII Newsletter* 6:14–16

Kala, C. P. 2004. Indigenous uses and structure of Chir Pine forest in Uttarakhand Himalaya, India. *International Journal of Sustainable Development and World Ecology* 11(2):205–210

Kala, C. P., and V. K. Uniyal. 1999. Forest vegetation along an altitudinal gradient in the Valley of Flowers National Park and its vicinity, Western Himalaya. *Annals of Forestry* 7(1):60–69

Kappelle, M., J. G. van Uffelen, and A. M. Cleef. 1995. Altitudinal zonation of montane *Quercus* forests along two transects in Chirripó National Park, Costa Rica. *Vegetatio* 119:119–153

Khan, J. A., A. Sultana, and M. S. Hussain. 2000. A study of threats to biodiversity conservation of middle altitude oak forest in Kumaun Himalaya. Final technical report submitted to Department of wildlife sciences, Aligarh Muslim University

Lovett, J. C. 1996. Elevational and latitudinal changes in tree associations and diversity in the Eastern Arc mountains of Tanzania. *Journal of Tropical Ecology* 12:629–650

Majila, B. S. 1992. Phytosociology, biomass structure and primary productivity of oak-pine forest of Kumaun Himalaya. [PhD thesis submitted to Department of Botany, Kumaun University, Nainital]

Majila, B. S., and C. P. Kala. 2010. Forest structure and regeneration along the altitudinal gradient in the Binsar Wildlife Sanctuary, Uttarakhand, India. *Russian Journal of Ecology* 41(1):75–83

Majila, B. S., G. C. Joshi, and C. P. Kala. 2005. Patterns in litter fall and litter decomposition along an altitudinal gradient in the Binsar Wildlife Sanctuary, Central Himalaya. *International Journal of Sustainable Development and World Ecology* 12(2):205–212

Monk, C. D. 1967. Tree species diversity in the Eastern deciduous forest with particular reference to North Central Florida. *American Naturalist* 101:173–187

Ralhan, P. K., A. K. Saxena, and J. S. Singh. 1982. Analysis of forest vegetation at and around Nainital in Kumaun Himalaya. *Proceedings of Indian National Science Academy* B 48:121–137

Rao, P. B. 1984. Regeneration of some trees of Western Kumaun Himalaya. [PhD thesis submitted to Department of Botany, Kumaun University, Nainital]

Risser, P. G., and E. L. Rice. 1971. Diversity in tree species in Oklahoma upland forests. *Ecology* 52:876–880

Saxena, A. K., and J. S. Singh. 1982. A phytosociological analysis of woody species in forest communities of a part of Kumaun Himalaya. *Vegetatio* 50:3–22

Shannon, C. E. and W. Weaver. 1949. *The Mathematical Theory of Communication.* Illinois: Urbana University Press

Simpson, E. H. 1949. Measurement of diversity. *Nature* 163:688

Singh, S. K. 1989. Studies on biomass, phytosociology and productivity of Siahi Devi Forest of West Almora division, Kumaun Himalaya. [PhD thesis submitted to Department of Botany, Kumaun University, Nainital]

Singh, J. S., and S. P. Singh. 1987. Forest vegetation of the Himalaya. *Botanical Review* 53:80–192

Uprety, N. 1982. A study on the phytosociology and state of regeneration of oak forest at Nainital. [PhD thesis submitted to Department of Botany, Kumaun University, Nainital]

Valdia, K. S. 1976. Kumaunkikhanizsampada: Kuchsambhawanayen. *Uttarakhand and Bhartiya* 1(2):1–6

Youngblood, A., T. Max, and K. Coe. 2004. Stand structure in east side old growth ponderosa pine forest of Oregon and northern California. *Forest Ecology and Management* 199 (2–3):191–217

CHAPTER 4

Population and Management of Medicinal Plants in Chhattisgarh

Chandra Prakash Kala

INTRODUCTION

Humans have been associated with the forest ecosystem since their appearance on Earth. Forests not only provide food, fuel, fodder, fibre, and shelter, but also their significance in overall environmental conservation is well recognized (Witt 1916; Singh 2002). Over the years, the plant species growing in various biogeographic zones and the types of forests have been studied for their medicinal properties, as evident from the historical records (CSIR 1989). Such records accumulated a great deal of knowledge on the use of plants for therapy; however the knowledge was restricted to a few specialized herbal healers in rural communities for a long time, due to which much of their use was observed as being of local interest. However, in the present century, the age-old traditional values attached with the various medicinal plants have gained tremendous importance and the major part of traditional herbal medicine, which includes folk health tradition, has transformed into trade (Kala 2005, 2006).

The current trend of increased commercialization of medicinal plants has resulted in overharvesting some economically important species. To understand and mitigate the consequences of overexploitation of medicinal plants and degradation of forest ecosystems, there is an urgent need to study the availability and regeneration patterns of useful species, especially of those with a high market demand. Additionally, the quantification of woody medicinal plants is an important aspect as they form the sources and habitat for many other faunal and floral species (Ronnells and Laumonier 2000; Sagar, Raghubanshi, and Singh 2003). This is further necessitated by the fact that the structure and composition of woody species affect human life in several interrelated ways including health security and livelihoods (Kala 2005; Kala, Sajwan, and Dhyani 2006).

The deciduous forest in the Chhattisgarh state of India contains a rich biological and cultural diversity, and represents the Deccan Peninsular biogeographic zone (Kala 2009). Many indigenous and local communities inhabit in these forests and meet their daily requirements from the surrounding forest resources. The effect of market-driven forces on the knowledge of medicinal plants and the population structure is inadequately explored. This is an important issue due to continuing commercial interest of communities and the shift of attitude towards the use of herbal medicine. The present study, therefore, deals with understanding the traditional use of medicinal plants and the population density of woody medicinal plants in the deciduous forest of Chhattisgarh. Attempts were also made to study the regeneration pattern of medicinally useful tree species with respect to the impact of ongoing anthropogenic and environmental activities.

STUDY AREA

The study was undertaken in the southern part of Surguja district that lies in the north of Chhattisgarh between 23°37'25" N to 24°6'17"N and 81°34'40" E to 84°4'40" E. The states of Uttar Pradesh, Jharkhand, Odisha, and Madhya Pradesh encircle Surguja district, and the Vindhyachal-Baghelkhand region of peninsular India overlaps south-eastern Surguja. Ambikapur is the district headquarters. The forests are dry, deciduous, and primarily dominated with *Shorea robusta*, *Madhuca indica*, *Anogeissus latifolia*, and *Semecarpus anacardium*. Many places of Surguja such as Ramgarh, Sita-Bhengra, and Laxmangarh have been associated with the famous Hindu epic *Ramayana* and named after Lord Rama, Laxmana, and Goddess Sita. Surguja is one of the important tribal regions of Chhattisgarh (Kumar 2007). The major ethnic groups in the study area were Gond, Majhwar, and Baiga. Apart from forest resource collection, the villagers practise agriculture and raise crops such as paddy and maize.

METHODOLOGY

Medicinal Plants Exploration

The field surveys were conducted from July to September 2008 in the tribal villages of Surguja, covering Kete, Ghatbarra, Parsa, Tara, Pendrakhi, Parogia, Hariharpur, Shivnagar, Fatepur, and Bhandargaun, to gather data on the therapeutic uses of plant species. During the survey period, information was also gathered using semi-structured questionnaires on the plant parts used for medicine and the types of ailments cured. The data was cross-checked with the help of group discussion held among different age groups of tribal and non-tribal villagers that included both genders of the society. The participant observation method was also employed to understand the methods and techniques adopted by the local people to use plants and plant parts for medicinal purposes. The forest areas around the villages were also surveyed accompanied by local youths and knowledgeable elders for the identification of species mentioned for curing ailments.

Medicinal Plants Population

For quantification of woody medicinal plants such as trees and shrubs, the nested plot technique along transects was used, following Misra (1968). In the forested areas, quadrats were laid down at 100 m intervals along with 1 km transects. The standard size of quadrat (10 m × 10 m for tree species and 5 m × 5 m for shrubs and saplings) was used for sampling. Plant species with >31.5 cm circumference at breast height (CBH) were considered as trees and those with <31.5 cm were considered as tree saplings. To understand the regeneration pattern of important woody medicinal plants, tree saplings were enumerated in 5 m × 5 m plots within each quadrat. The trees, shrubs, and saplings with their individuals were counted in each quadrat. A total of eight such transects were laid down in the forests.

The collected data was analysed for frequency, density, and basal area by using the following formulae as given by Misra (1968).

$$\text{Frequency} = \frac{\text{Number of quadrats in which species occur}}{\text{Total number of quadrats}}$$

$$\text{Density} = \frac{\text{Total number of individuals}}{\text{Total number of quadrats}}$$

Basal area = πr^2 where r is the radius

RESULTS AND DISCUSSION

Social Dimension of Medicinal Plants

The present study has resulted in the documentation of 36 medicinal plant species used by tribal and non-tribal communities to cure various diseases (Table 1). The documented plants were distributed over 26 families, of which Combretaceae and Liliaceae represented the highest number of medicinal plant species. Of these, 14 were herbaceous species, 13 were trees, 6 were woody climbers, and 3 weres shrubs (Figure 1). Different plant parts and extracts of these species, such as roots, tubers, leaves, fruits, barks, resins, seeds, and latex were used as medicine. In the majority of the cases, roots (14 species) were used to prepare medicine, followed by fruits (7 species) and bark (5 species). More than one plant part of the four plant species such as *Garuga pinnata* Roxb., *Helicterus isora* L., *H. antidysenterica* Wall., and *Ficus bengalensis* L. were used as medicine. Cough, bodyache, dysentery, cut-wounds, scorpionbite, snakebite, muscular pain, and indigestion were the major ailments cured by these plant species.

The study area harboured several important medicinal plants used in Ayurvedic medicine. *Terminalia arjuna* Bedd., *Terminalia*

Table 1 *Plant species used by local people in the study area to cure ailments*

Botanical name	*Family*	*Local name*	*Habit*	*Use*	*Part used*
Achyranthus aspera L.	Amaranthaceae	Gathiya, aghada	Herb	Diuretic, tonic, insect and scorpion bites	Leaf
Acorus calamus L.	Acoraceae	Bach	Herb	Stomach ache, anthelmintic	Root
Aegle marmelos Correa.	Rutaceae	Bel	Tree	Dysentery	Fruit
Asparagus racemosus Willd.	Liliaceae	Kargi	Shrub	Internal pain, tumours, tonic	Root
Boswellia serrata Roxb.	Burseraceae	Saliha	Tree	Arthritis	Resin
Caesalpinia bonducella Flem.	Caesalpiniaceae	Gataran	Woody climber	Tonic	Seed
Celastrus paniculata Willd.	Celastraceae	Unjain	Woody climber	Wounds	Seed
Ceropegia bulbosa Roxb.	Asclepiadaceae	Bosiya kandha	Climber	Tonic, aphrodisiac	Root
Chlorophytum tuberosum Baker	Liliaceae	Safed musli	Herb	Skin diseases, tonic	Tuber
Chloroxylon swietenia DC.	Meliaceae	Bhirra, bharahi	Tree	Mosquito repellent	Leaf
Cissus quadrangularis L.	Vitaceae	Hathjod	Climber	Antimicrobial, fracture, fertility	Root
Curcuma angustifolia Roxb.	Zingiberaceae	Tikhur	Herb	Wounds	Root
Cynodon dactylon Pers.	Poaceae	Doob	Grass	Liver disorder, wounds	Root
Dioscorea bulbifera L.	Dioscoraceae	Agitha	Climber	Tonic, tumour	Tuber
Dioscorea sp.	Dioscoraceae	Gethi kandha/ Karuha	Herb	Tonic, tumour	Fruit
Diospyros melanoxylon Roxb.	Ebenaceae	Tendu	Tree	Snakebite	Root
Elaeodendron glaucum Pers.	Celastraceae	Mamri, Jamrasi	Shrub	Snakebite	Root

Contd...

Table 1 *Contd...*

Botanical name	*Family*	*Local name*	*Habit*	*Use*	*Part used*
Ficus bengalensis L.	Moraceae	Gad nifir	Tree	Dysentery	Latex, bark
Garuga pinnata Roxb.	Burseraceae	Khenkara, Kekad	Tree	Snakebite, wounds	Bark, fruit
Gloriosa superba L.	Liliaceae	Karihari	Herb	Tumors, skin diseases	Root
Helicterus isora L.	Sterculiaceae	Aaithi, marorphali	Shrub	Colic, intestinal disorder	Bark, fruit
Hibiscus abelmoschus L.	Malvaceae	Kapalsiya kandha	Herb	Blood in urine	Root
Holarrhena antidysenterica Wall.	Apocynaceae	Koriya	Tree	Fever, dysentery	Root, Bark
Ipomoea mauritiana Jacq.	Convolvulaceae	Patal kohra	Woody climber	Indigestion	Root
Peucedanum nagpurense (Cl.) Prain	Asteracease	Tejraj	Herb	Medicine	Root
Phoenix acaulis Buch	Palmaceae	Chind	Herb	In child birth	Fruit
Phyllanthus emblica L.	Euphorbiaceae	Awala	Tree	Indigestion	Fruit
Semecarpus anacardium L.	Anacardiaceae	Bhelwa	Tree	Infection	Fruit
Soymida fabrifuga A. Juss.	Meliaceae	Rohina, rohan	Tree	Muscular pain	Bark
Terminalia arjuna Bedd.	Combretaceae	Kahua, arjun	Tree	Heart disorder, fever, asthma,	diarrhea Bark
Terminalia bellerica Roxb.	Combretaceae	Baira	Tree	Cough	Fruit
Terminalia chebula Retz.	Combretaceae	Harra	Tree	Cough	Fruit
Urginea indica Kunth.	Hyacinthaceae	Ban pyaz	Herb	Scorpion bite	Tuber
Vallaris heynei Spr.	Apocynaceae	Dudhiya kandha	Woody climber	Lactating mother	Latex
Vitis carnosa Wall.	Vitaceae	Dhokar bela	Woody climber	Bodyache	Root
Zizyphus rugosa Lamk.	Rhamnaceae	Churaban, churna	Woody climber	Bodyache	Whole

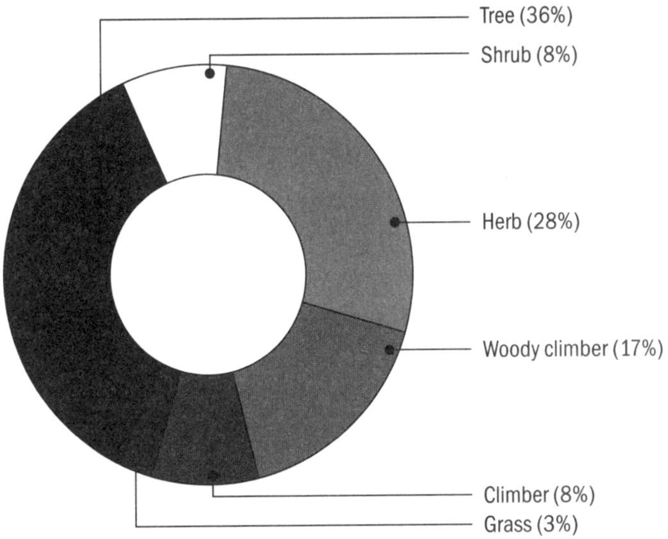

Figure 1 *Distribution of medicinal plant species across various life forms in the study area*

bellerica Roxb., *Terminalia chebula* Retz., *Phyllanthus emblica* L., *H. antidysenterica* Wall., *Hibiscus abelmoschus* L., *Gloriosa superba* L., *Dioscorea bulbifera* L., *Aegle marmelos* Correa., *Boswellia serrata* Roxb., *Acorus calamus* L., and *Asparagus racemosus* Willd. were among the prominent ones, as listed in the ayurvedic medicine. Three important species of Ayurvedic medicine, that is, *T. arjuna*, *T. bellerica*, and *P. emblica*, generally known together as *Triphala*, were available and used by local people in the study area. *Celastrus paniculata* is one of the precious medicinal herbs valued for its use in Ayurveda to sharpen the memory and increase intellect.

Snakebites are a common problem in the study area as it harbours many species of poisonous snakes. Snakebites were treated by an expert who used plants, but the use of such species was not disclosed to the common people due to several reasons including conservation of species from overexploitation. The importance of *Diospyros melanoxylon* Roxb., *Elaeodendron glaucum* Pers., and *G. pinnata* is very high due to their use in curing snakebites. Similarly, scorpionbites were treated using a paste of *Achyranthus aspera* L.

leaves and the tuber of *Urginea indica* Kunth. The Surguja district is also a mosquito prone area, and incidence of malarial fever is common. The local people spend much of their time in the forest for rearing livestock and collecting fuelwood, fodder, and medicinal and edible plants. To keep the mosquitoes away, they rub the leaves of *Chloroxylon swietenia* DC on the exposed parts of the body and also carry its twigs on the head and back.

The traditional practice of harvesting medicinal plants is being invaded by modern commercial forces. Earlier, the collection of useful species from the wild was done by skilled people who knew how to harvest them. Today, in the rush to earn cash, unskilled labourers from outside the community are also harvesting medicinal plants. The attitude of the local people has also changed with the introduction of cash in medicinal plant collection. Altogether, this has led to overexploitation of many important medicinal plant species in the wild, and damaging the natural population of the respective species. In the nearby district of Bilaspur, 60% of the *P. emblica* trees have been damaged due to faulty harvesting practices (Tiwari 1995). The problem of overexploitation of resources needs to be handled properly, considering the livelihood and health security of the local people.

Ecological Dimension: Distribution and Population Status

There was a high heterogeneity in the distribution pattern of woody medicinal plant species in the study area. Of the total 16 woody medicinal plant species (13 trees and 3 shrubs), 12 occurred in the sampling plots of the field survey. Of these, nine were tree species and three were shrubs. Among all the tree species used as medicine, *D. melanoxylon* Roxb. had the highest density (32.14 individuals/ha) and frequency whereas *T. bellerica* Roxb. and *F. bengalensis* L. had the lowest density and frequency in the study area. The distribution pattern and population density of *B. serrata* Roxb. and *P. emblica* L. were similar (Table 2).

The basal area of medicinal tree species varied from species to species. *F. bengalensis* L. had the highest basal area (2.76 per m^2/ha), followed by *D. melanoxylon* Roxb. (1.93 per m^2/ha), and *B. serrata* Roxb. (1.47 per m^2/ha). *T. bellerica* Roxb. and *G. pinnata* Roxb. had the lowest basal

Table 2 *Frequency, density, and basal area of ethno-medicinal tree species, shrub species, and tree saplings in the study area*

Species	*Local name*	*Frequency*	*Density (per ha)*	*Basal area (m^2/ha)*
Tree species				
Boswellia serrata Roxb.	Saliha	0.10	9.52	1.47
Diospyros melanoxylon Roxb.	Tendu	0.29	32.14	1.93
Ficus bengalensis L.	Bargad	0.01	1.19	2.76
Garuga pinnata Roxb.	Kenkara	0.05	5.95	0.14
Phyllanthus emblica L.	Awala	0.08	9.52	0.38
Semecarpus anacardium L	Bhelwa	0.05	4.76	0.31
Terminalia bellerica Roxb.	Baira	0.01	1.19	0.09
Terminalia chebula Retz.	Harra	0.02	2.38	0.16
Tree sapling				
Diospyros melanoxylon Roxb.	Tendu	0.85	6683.10	-
Garuga pinnata Roxb.	Kekad	0.11	63.38	-
Holarrhena antidysenterica Wall.	Korya	0.20	323.94	-
Phyllanthus emblica L.	Awala	0.02	21.13	-
Semecarpus anacardium L.	Bhelwa	0.08	56.34	-
Terminalia chebula Retz.	Harra	0.02	21.13	-
Shrub				
Asparagus racemosus Willd.	Asparagus	0.02	14.08	-
Elaeodendron glaucum Pers.	Jamrasi	0.07	49.30	-
Helicterus isora L.	Aeithi	0.02	14.08	-

area among all the medicinal tree species. Although all the three medicinal shrub species mentioned by the local people were found in the sampling plots during the field investigations, the frequency and density of these species were quite low. *E. glaucum* Pers. had a relatively high density (49.30 individuals per ha) compared to *A. racemosus* Willd. and *H. isora* L.

The regeneration pattern in medicinal tree species was uncertain to a large extent. In the sampling plots, three important medicinal tree species *B. serrata* Roxb., *T. bellerica* Roxb., and *F. bengalensis* L. were not encountered in the sapling stage. Similarly, *H. antidysenterica* Wall. was only found in the sapling stage and no mature individual of this tree species was encountered in the sampling plots. *D. melanoxylon* Roxb. had the highest density (6683 saplings/ha) and frequency (0.85) of saplings in the study area, followed by *H. antidysenterica* Wall. (Table 2). Natural disturbances as well as anthropogenic activities have influenced the growth of medicinal plants in natural habitats. The selective felling of medicinal plants, coupled with destructive harvesting practices, maybe the major cause of concern for the present composition and structure of medicinal plants in the forests.

Rarity and Management Concerns

Four important medicinal tree species *A. marmelos* Correa, *Chioroxylon swietenia* DC., *Soymida fabrifuga* A. Juss, and *T. arjuna* Bedd. did not occur in the sampling plots, which indicate their low population in the study area. The study area also harboured some threatened medicinal plants of the IUCN Red List categories. These species were *A. calamus* L., *G. superba* L., *D. bulbifera* L., *Curcuma angustifolia* Roxb., *Pterocarpus marsupium* Roxb., *Peucedanum nagpurense* (Cl.) Prain, *Costus speciosus* (Koen.) Sm., *Chlorophytum tuberosum* Baker, and *C. paniculata* Willd. These species have become threatened due to several reasons, including their repeated harvest and over-collection for commercial use.

The reason why some species are becoming endangered varies from species to species. *A. calamus* L. has become endangered due to habitat specificity, as it was found in the marshy and waterlogged areas, which are declining, and over the past decade its population has declined to upto 80% in Chhattisgarh (Ved, Kinhal, Ravikumar, *et al.* 2003). *A. calamus* is used for medicine by the different ethnic communities in the area. The large demand for its rhizomes in the local as well as international market has accelerated its exploitation in the forest. The peculiar habitat of *T. arjuna* Bedd., mainly found along water courses in the study area, was one of the reasons for its low population. A similar phenomenon occurred with *C. angustifolia* Roxb., which mostly grows in moist hillock areas. *C. paniculata* Willd. is a woody climber; the removal of this climber for seeds has caused its population to decline due to poor regeneration. As *C. paniculata* Willd. does not grow easily, there is difficulty in the propagation and cultivation of this species. Due to early harvesting of the rhizomes of

C. speciosus (Koen.) Sm., seed formation was severely affected and, thus, created problems for its regeneration.

The gum of *B. serrata* Roxb. is used as medicine, so its extraction with destructive harvesting practices imposes a major threat on its population in the wild. *B. serrata* Roxb. is one of the highly traded medicinal plant species (Ved and Goraya 2007). The extensive use of its wood for packing and plywood is leading to further depletion of its population in the wild. The regeneration capacity of this species is poor, as is evident from the present study—not a single sapling was encountered in the sampling plots, and therefore requires special care and management. Roots and seeds of *P. nagpurense* (Cl.) Prain are collected for trade and this herb, too, was not found in the sampling plots due to low population. Similarly, destructive harvesting of the fruits of *P. emblica* L., by cutting the entire fruit-bearing twigs, was one of the major causes of concern when conserving this species. The underground plant parts such as tubers of *C. tuberosum* Baker are collected. Apart from medicine, this species is being used in the carpet and tobacco industries, creating pressure on the wild population.

All threatened medicinal plants are important due to their unique health-care properties and, hence, require special conservation measures. The medicinal tree species need to be restored through plantations in their natural habitat range. The shrub species may be grown as the under story species with the tree plantations. Species like *A. calamus* that occurs in the waterlogged and marshy areas may be planted in the bunds of ponds where marshy conditions exist. Ex-situ conservation through traditional and modern tools such as the use of rural biotechnology should be done on a large scale for the survival of these species as well as for the health security of the community.

Traditional management norms in the form of sacred groves, sacred species, and sacred landscape may need proper respect in conservation planning. Some of the localities in the present study area, such as Ramgarh, are examples of a traditional conservation doctrine through which the genetic diversity of many valuable medicinal plants are being conserved. Since Ramgarh is a sacred land for the local people devoted to Lord Rama, the overexploitation of any natural resource is strictly prohibited here. The medicinal plants are very important to the local people residing in and around the forests, which is why such produce should be protected from overexploitation and degradation. This may be accomplished by designing non-destructive harvesting methods. Rather than supplying medicinal plants in the raw form, efforts should be made to promote processing and value addition at the local level.

Compiling a list of indigenous uses of the medicinal plants was not an easy task as most native healers were reluctant to accurately share their knowledge with outsiders. The documented medicinal plants may play an important role in scientific research. Also, it is believed that herbal medicines are much safer than allopathic medicines due to few or no side effects. Many medicinal plants described in this study have a high potential for wide application in the pharmaceutical industry.

ACKNOWLEDGEMENTS

The author would like to thank the Director, Indian Institute of Forest Management (IIFM), Bhopal for his support. During the field surveys, the help rendered by the fieldstaff, especially Ramdhan, Syam Nath, and Sukh Nandan, is greatly appreciated.

REFERENCES

CSIR (Council of Scientific and Industrial Research). 1989. *The Wealth of India: Raw Materials.* New Delhi: Publication and Information Directorate, CSIR

Kala, C. P. 2005. Indigenous uses, population density, and conservation of threatened medicinal plants in protected areas of the Indian Himalayas. *Conservation Biology* 19(2):368–378

Kala, C. P. 2006. Ethnobotany and ethnoconservation of *Aegle marmelos* (L.) Correa. *Indian Journal of Traditional Knowledge* 5(4):537–540

Kala, C. P. 2009. Aboriginal uses and management of ethnobotanical species in deciduous forests of Chhattisgarh state in India. *Journal of Ethnobiology and Ethnomedicine* 5:1–12. <Details available at www.ethnobiomed.com/content/5/1/20>

Kala, C. P., B. S. Sajwan, and P. P. Dhyani. 2006. Developing the medicinal plants sector in northern India: challenges and opportunities. *Journal of Ethnobiology and Ethnomedicine* 2:1–15

Kumar, K. 2007. Working plan of North Surguja, Ambikapur. Chhattisgarh: Forest Department, Government of Chhattisgarh

Misra, R. 1968. *Ecology Work Book.* New Delhi: Oxford and IBH Co.

Ronnells, K., and Y. Laumonier. 2000. Species diversity structure analysis at two sites in the tropical rain forest of Sumatra. *Journal of Tropical Ecology* 16:253–270

Sagar, R., A. S. Raghubanshi, and J. S. Singh. 2003. Tree species composition, dispersion and diversity along a disturbance gradient in a dry tropical forest region of India. *Forest Ecology and Management* 186(1–3):61–71

Singh, J. S. 2002. The biodiversity crisis: a multifaceted review. *Current Science* 82:638–647

Tiwari, K. P. 1995. Collection of Aonla (*Emblica officinalis*) fruits from forest: an impact assessment. *Vaniki Sandesh* XIX(4):1–5

Ved, D. K., G. A. Kinhal, K. Ravikumar, M. Karnat, R. Vijaysankar, and J. H. Indresha. 2003. *Threat Assessment and Management Prioritization for the Medicinal Plants of Chhattisgarh and Madhya Pradesh.* Bengaluru: Foundation for Revitalization of Local Health Traditions

Ved, D. K., and G. S. Goraya. 2007. *The Demand and Supply of Medicinal Plants in India.* Bengaluru: Foundation for Revitalization of Local Health Traditions

Witt, D. O. 1916. *Descriptive List of Trees, Shrubs, Climbers and Economic Herbs of the Northern and Berar Forest Circles Central Provinces.* Allahabad: Pioneer Press

CHAPTER 5

Application of Remote Sensing and GIS in Mapping the Plant Resources of Odisha

Chiranjibi Pattanaik, C Sudhakar Reddy, and P S Roy

INTRODUCTION

Biodiversity refers to the variety of living organisms at all levels of organization, from species through gene, to higher taxonomic levels, including the variety of habitats and ecosystems, as well as the processes occurring therein. The levels of biodiversity can be expressed in a hierarchical structure (Figure 1) with the genetic, population-species, community, ecosystem, and landscape. Each of these levels can be further divided into compositional, structural, and functional components. Composition includes the genetic constitution of population, the identity and relative abundance of species in a natural community, and the kinds of habitats and communities distributed across the landscape. The structure includes vertical layering and horizontal patchiness of vegetation. Function includes the climatic, geologic, hydrologic, ecological, and evolutionary processes that generate biodiversity and keep it forever changing

Though biodiversity is generally appreciated at the species level, it needs to be assessed and conserved at all levels of ecological organization and spatio-temporal scales (Figure 2). These include

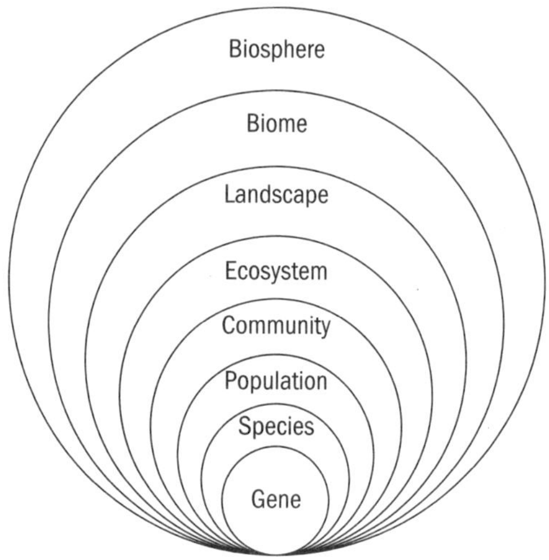

Figure 1 *Hierarchical depiction of biological organization in nature (DBT–DOS biodiversity manual)*

ecosystems containing rich biodiversity, large number of threatened or endemic species that are important for migrating species, those that have social, economic, cultural, or scientific significance, or support key processes, species and communities of species that are threatened, related to domesticated or cultivated species, have medicinal, agricultural, or other economic, social, cultural, or scientific significance, and indicator species, genotypes with social, scientific, or economic significance. There is a possibility that the species loss may even overtake the inventorying efforts with the conventional species level approach. A more acceptable approach suggests the conservation of habitats rather than a single species.

Global biodiversity assessment (Heywood 1995) estimates the total number of animal and plant species in the world to be about 14 million. It further records that, so far, only 1.75 million species have been described and studied. The United Nations Environment Programme's (UNEP) World Conservation Monitoring Centre (WCMC) estimates around 270 000 species of vascular plants and 52 000 animals (vertebrates) (UNEP-WCMC 2000). Nearly 45% of the world's vascular plant species occur in closed tropical forests of the world (Myers, Mittermeier, Mittermeier, *et al.* 2000).

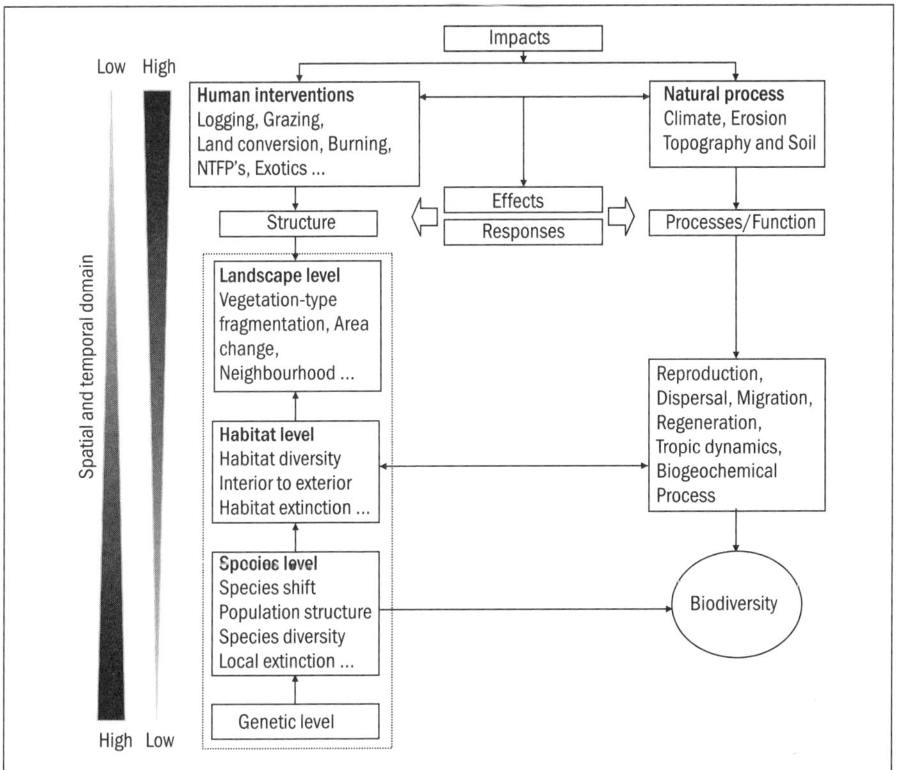

Figure 2 *Components of biodiversity assessment (DBT–DOS biodiversity manual)*

Vegetation in the Indian sub-continent is distributed mainly in four distinct geographical zones—the Himalayas, the Vindhyas, and the Western and Eastern Ghats. India possesses a rich diversity of flowering plants (17 000 species) with a high degree of endemism (33.5%). The Western Ghats, one of the biodiversity hotspots of the world, alone contain 4000 species of flowering plants, of which 1500 species are endemic (Nayar 1989; Myers, Mittermeier, Mittermeier, *et al.* 2000). Nayar (1996) has recognized 40 hotspots of Indian flora, with concentration of endemics as a major criterion. Similarly, the Eastern Ghats have nearly 3200 flowering plants, of which 136 species are endemic (Reddy, Brahmam, and Raju 2006). With varied agro-climatic conditions, diversified edaphic factors, and different altitudinal ranges, Odisha has built enormous genetic variation in its natural flora. There are 2727 species of vascular plants found in

Odisha (Saxena and Brahmam 1996). The endemic flora of the state is represented only by 28 taxa (Saxena and Brahmam 1983).

All global conservation assessments—whether focused on the coverage of protected areas (Rodrigues, Andelman, Baker, *et al.* 2004), the impacts of habitat loss (Gaston, Blackburn, and Goldewijk 2003), or the potential effects of climate change (Thomas, Cameron, Green, *et al.* 2004)—require information on the spatial distribution of the elements of biodiversity. This knowledge of the biodiversity pattern provides the essential foundation on which more sophisticated assessments of ecological, evolutionary, and socio-economic processes can be buit. Changing land cover patterns and growing population are largely responsible for the rapid loss of biodiversity. Under the current scenario, about 20% of all species are expected to be lost within 30 years and 50% or more by the end of the 21st century (Myers 1993). Global concern over the loss of biodiversity has forced land managers to evolve better methods for conserving it at various spatial and temporal scales and the same concern has been further emphasized by the Convention on Biological Diversity (CBD). In an important study, high correlation between threat rates and percentage of endemic species would seem attributable to the fact that a narrow geographic range makes species more vulnerable to extinction (McKinney 1997; Purvis, Gittleman, Cowlishaw, *et al.* 2000).

The site-specific studies use theoretical principles in conservation biology that are often inadequately tested in the field. These principles are being used as a framework to guide management plans, with obvious limitations of being applicable in the spatial context (Abbott 1980). Application of landscape ecological principles for prioritizing biodiversity-rich sites has the advantage of integrating spatial information, non-spatial information, and horizontal relationships in space and time. The landscape approach for biodiversity characterization also addresses some of the limitations of ground-based point inventory.

Recent developments in satellite remote sensing and a Geographic Information System (GIS) coupled with user-oriented computer programs allow us to use landscape ecological principles for biodiversity characterization at landscape level more efficiently (Roy, Murthy, Singh, *et al.* 2008). The present study hypothesizes biological richness at landscape level as a function of habitat, biogeographical setting, disturbance regime, and environmental complexity. The article presents an approach for biodiversity characterization at landscape level using geospatial techniques. The approach has been validated in one of the biologically rich sites of Odisha. Remote sensing data has been used primarily to stratify habitats, vegetation types, land use, and different land cover association. The ground data by stratum has provided properties of the habitats. Environmental complexity can be spatially estimated based on terrain and climate variability. The surrogate parameters determining phytodiversity have been taken into consideration for the geospatial modelling of biodiversity characterization at landscape level.

STUDY AREA

The state of Odisha is situated on the eastern coast of the Indian peninsula, between 81°24' E to 87°29' E and 17°48' N to 22°34' N. Bounded on the north by Jharkhand, on the west by Chhattisgarh, on the south by Andhra Pradesh, on the north-east by West Bengal, and on the south-east by Bay of Bengal with a coastline of 482 km, the state covers an area of 155 707 km^2 (Figure 3).

Odisha has a rich forest area, mainly dominated by sal mixed types. The north-western part of the state forms part of the Chhotanagpur Plateau and has vegetation mostly dominated by sal forests, while the central and eastern regions of the Eastern Ghats are part of the central highlands of the Deccan Peninsula, and are dominated by mixed deciduous forests. The north-eastern and southern parts have mountainous terrain where the Eastern Ghats run parallel to the coast from Mayurbhanj to Malkangiri districts. This region is biologically very rich, harbouring a large number of economically important and medicinal plants unique to the region. The north-south chain of mountains also exhibit gradients in aspect and altitude. The altitudinal gradient, coupled with the southwest and northeast rainfall patterns and soil conditions, has resulted in a

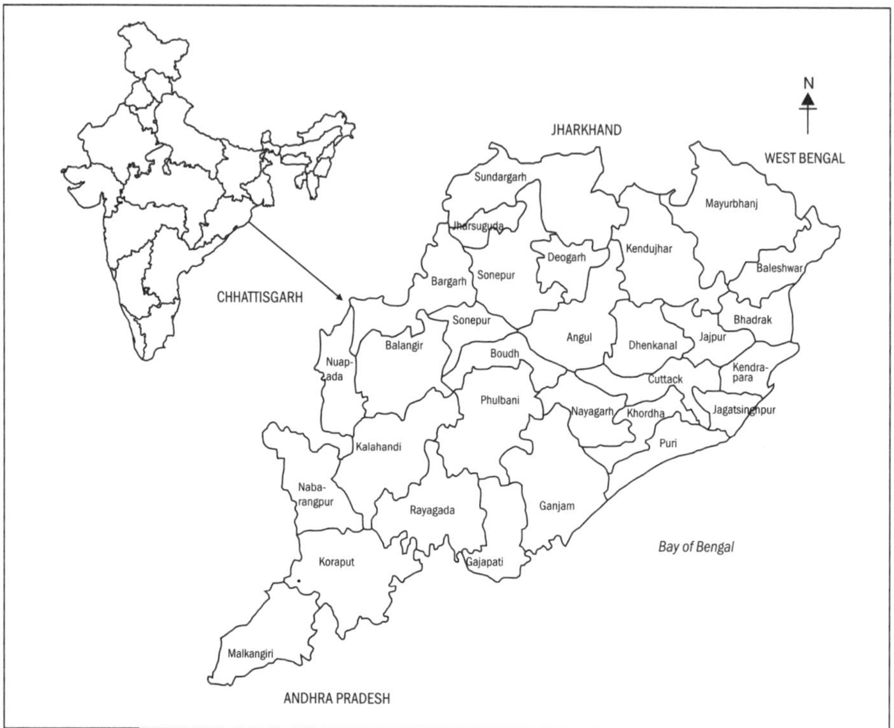

Figure 3 *Location map of Odisha with all districts*

high degree of habitat heterogeneity. Further, the seasonal variations in the form of dry and moist months influence the phenological types representing great biological diversity.

The temperature ranges from 12°C to 42°C in the study area. It begins to rise early in March and continues to rise till the end of May or early June, when it reaches about 42°C in most areas (Koraput, Cuttack, and Kalahandi districts) except the hills. May is the hottest month and December is the coldest. In the Daringibadi area of Phulbani district, the temperature drops as low as 0.5°C and causes occasional snowfall. The normal annual rainfall in the study area is 900 mm, but rises to 2000 mm in the hills. Odisha ranks fourth amongst state/union territories of the country in terms of recorded forest area (58 135 km^2) and present forest cover (48 366 km^2). The forested area is 31.06% of the total geographic area of the state and 7.16% of the total forest cover of the country (FSI 2003).

MATERIALS AND METHODOLOGY

Remote Sensing and Ancillary Data

The Indian Remote Sensing (IRS) satellite RESOURCESAT-1 Linear Imaging Self Scanner (LISS) III datasets were procured from NRSC Data Centre (NDC), National Remote Sensing Centre. The sensor provides data in four bands: two visible (green: 0.52–0.59 μ and red: 0.62–68 μ), one near infrared (NIR: 0.77–0.86 μ), and one in short wave infrared (SWIR: 1.55–1.70 μ) with a spatial resolution of 23.5 m. The data used was acquired between November 2003 and January 2004. In addition to this, topographic maps, forest management maps, and protected area maps were used to generate the ancillary layers and ground work.

Satellite Data Processing

The digital data of IRS P6 LISS III was initially loaded into the computer hard disk using the ERDAS Imagine 8.6 image processing software. Radiometric correction was carried out over the raw data sets using the first-order dark pixel subtraction technique. Geometric correction was done using topographic maps as reference. The identification of Geometric Control Points (GCPs) on the satellite image and the corresponding satellite image position were carefully selected and assigned in a well-distributed fashion across the entire image set. All the images were rectified with UTM projection and WGS84 datum. A False Colour Composite (FCC) image of Odisha was generated (Figure 4).

Visual Interpretation Technique

A template depicting different vegetation types and certain class codes need to be prepared and followed for visual interpretation. The visual interpretation technique was followed, using certain fundamental image characteristics that help interpret earth features. These are tone, texture, pattern, size, shape, and shadow, coupled with site/location and associated features. After this, the satellite data was interpreted on screen on a 1:50 000 scale in polygon mode, using ARC GIS 9.0 software. Immediately after the closure of each

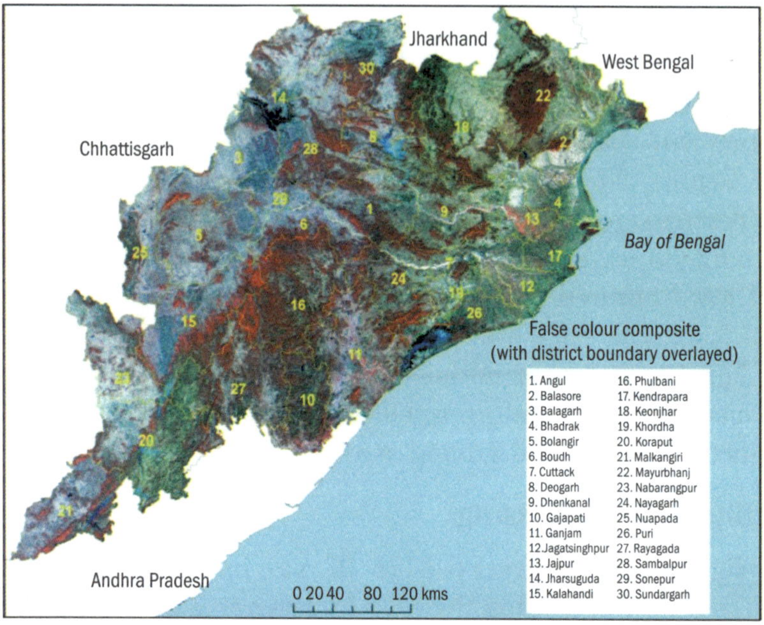

Figure 4 *False colour composite (FCC) of IRS P6 LISS III image of Odisha*

polygon, pre-decided class assignment orpolygon IDs were attributed. At the end of image interpretation, area statistics were obtained.

Ground Truthing

The field survey was undertaken to get acquainted with the general patterns of vegetation and habitat types in the area. Major cover types and a few prime localities of the characteristic types were recorded. The image elements were correlated with the ground truth realities and the interpretation key developed. The variation and tonal patterns were observed on existing maps/images. Major roads, drainage, and hill tops were traversed to collect ground information. A thorough literature survey and interaction with local institutions were also made for the present study.

Biological Richness Modelling

The regional vegetation, physiography, land use, and climate determine landscape characteristics. A vegetation cover, or land use-type map, is the prime input for a landscape analysis to derive intermediate layers. This information, along with phytosociological data, bioclimate, and topography was used for biodiversity characterization. The geospatial model used at present for Biodiversity Characterization at Landscape Level (BCLL) is shown in Figure 5.

RESULTS AND DISCUSSION

Phytosociological Analysis

Phytosociological analysis was carried out to understand the floristic and vegetation pattern in the state. For the phytosociological analysis,

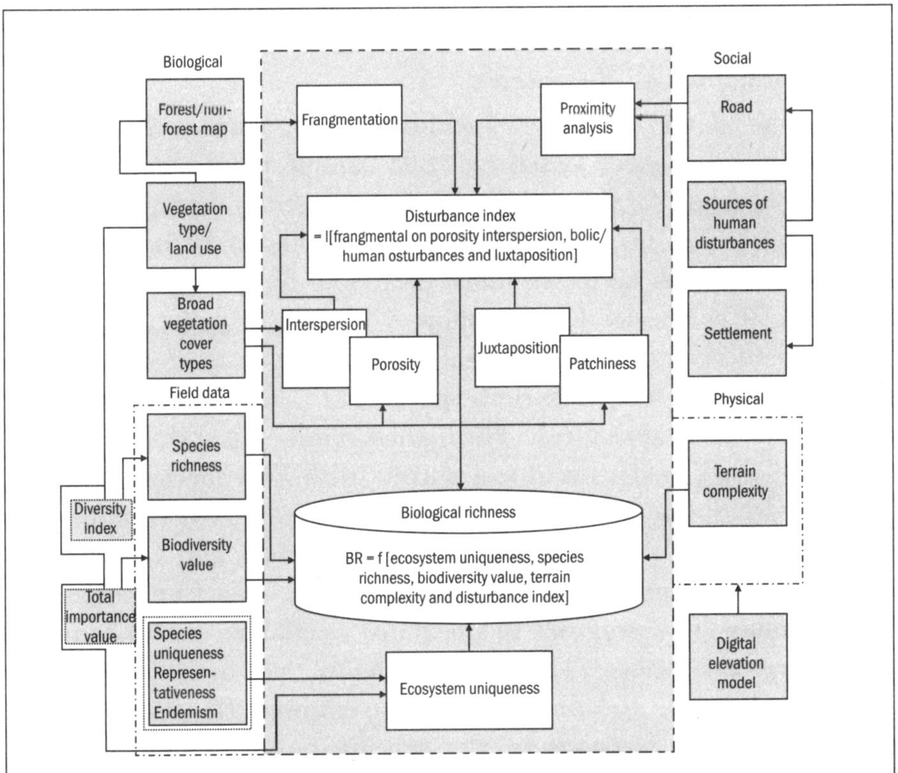

Figure 5 *Schematic representation of the geospatial model for BLCC using spatial landscape analysis modelling (SPLAM)*

the vegetation types were grouped into five major types. The forest cover of Odisha has a spatial extent of 48 838 km^2. In order to carry out the ground-level phytosociological studies, a sampling intensity of 0.002% of forest area was envisaged. Accordingly, the entire state required 2400 sample points, distributed proportionally among all the major forest types. However, based on accessibility and local law and order problems, 1803 sample points were proportionally covered with 113, 417, 750, 386, 97, and 40 sample points for semi-evergreen, moist deciduous, sal mixed moist deciduous, sal mixed dry deciduous, dry deciduous, and mangroves, respectively. Based on the species area, curve developed initially and studying the land covers of the region, a plot size of 0.04 ha was adopted uniformly for all the types. In each plot, names of the trees (with GBH and height), herbs, shrubs, climbers, seedlings, and saplings were recorded for all the species.

A total of 1173 species, including trees, shrubs, herbs, and climbers were recorded based on 1803 sample points (each sample plot was 0.04 ha). Of these, 349 species belonged to the tree category, followed by 172 species of shrubs, 510 species of herbs, and 142 species of climbers. sal mixed moist deciduous forests had the highest number of tree species (241), followed by 203 in moist deciduous, 199 in semi-evergreen, 154 in sal dry deciduous, and 137 in dry deciduous forests. The most herb species (317) were found in the sal mixed moist deciduous forest. The highest number of species has been recorded for sal moist deciduous forests with 785 species, followed by moist deciduous type with 625 species. The sal dry deciduous, dry deciduous and semi-evergreen types had 442, 347, and 484 species, respectively. The maximum diversity observed, based on Shannon–Wiener index (H'), was 6.2 in the moist deciduous and 6.1 in the semi-evergreen forests. The dry deciduous, sal moist deciduous, sal dry deciduous, and mangrove forests showed H' values of 5.8, 4.8, 4.4, and 3.4, respectively. The species similarity index between vegetation types was calculated and the highest similarity was found between moist deciduous and dry deciduous types, followed by sal moist deciduous and moist deciduous forests.

A total of 14 endemic/threatened species were recorded during phytosociological sampling in Odisha. The identification of endemics is based on the available literature (Saxena and Brahmam 1996; Reddy, Reddy, and Jadav 2001). The highest number of endemics, 12 species, has been recorded in sal mixed moist deciduous forests. Of these, eight species are included in the Red Data Book of Indian Plants (seven vulnerable and one endangered). The Total Importance Value (TIV) for each species was calculated considering 10 important uses (Ambastha 1992). The parameters considered are as follows.

- Grazing
- Medicinal
- Food
- Timber
- Charcoal
- Dye
- Fuel
- Oil
- Tannin
- Others

These parameters have been weighted on a scale of 1 to 10 where the maximum value represents the highest economic value. The maximum TIV was observed in the dry deciduous type (TIV = 10.81) while the semi-evergreen type had the least value (TIV = 6.25). A list of 374 economically important species and 228 medicinal plants were recorded with the help of available literature (Ambastha 1992; FRLHT 1998) in Odisha.

Landscape Mapping

The landscape was mapped in terms of fragmentation, patchiness, interspersion, porosity, and juxtaposition, using a spatial analysis of the vegetation-type map generated using remote sensing data. Of the total geographical area, around 31% is under forest cover, calculated from the vegetation-type map of the study area (Figure 6). Of the non-forest area, 44% is under agriculture. Scrub-land and barren areas constitute about 8.6% and 4.7%, respectively. Areas covered by water bodies in the state are around 3.3%.

The semi-evergreen forests are the least fragmented, with 89.2% in the low fragmentation category, followed by sal mixed moist deciduous

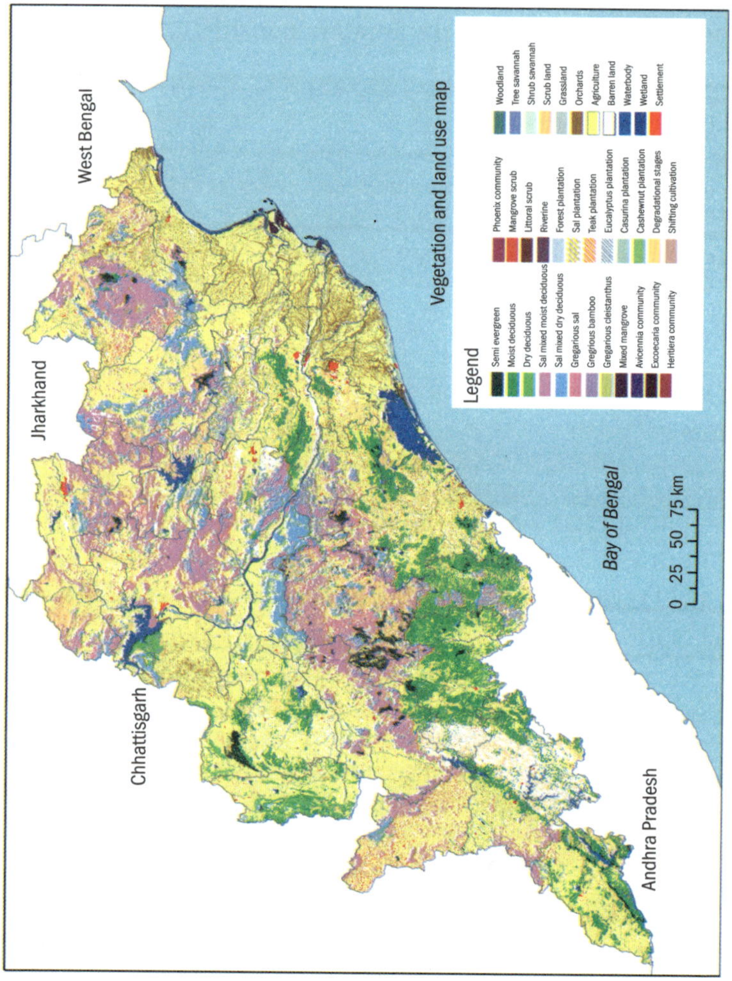

Figure 6 *Vegetation and land use map of Odisha*

forests, 76.1%, and moist deciduous forests, 73.6%. The sal mixed dry deciduous forest (2.5%) and the dry deciduous forests (2.3%) come under the high fragmentation category. This indicates that the remnant forests in Odisha still have undisturbed core forests with little or no anthropogenic disturbances. The disturbance index map is generated by integrating various landscape parameters of human disturbance, porosity, patchiness, juxtaposition, and interspersion. The medium disturbed areas are high (40.4%) in dry deciduous forests, followed by 31.9% and 29.7% in sal dry deciduous forests and moist deciduous forests, respectively. The low disturbed areas were found high in the semi-evergreen forests (73.7%), followed by sal mixed moist deciduous (69.2%), moist deciduous (61.8%), and sal dry deciduous forests (58.7%).

The biological richness (BR) map of Odisha (Figure 7) is a product of disturbance index, terrain complexity, ecosystem uniqueness, species richness, and biological value and is divided into low, medium, high, and very high categories. The overall distribution of BR in Odisha shows maximum BR at the medium level (44%), followed by low (37%), high (16%), and very high (3%). Of the different forest types in Odisha, the semi-evergreen forests represent very high BR, followed by moist deciduous, dry deciduous, sal mixed moist deciduous, and sal mixed dry deciduous, respectively. The high BR areas are dominated by semi-evergreen and moist deciduous forests. The very high biodiversity zones have been found in Similipal Biosphere Reserve, Gandhamardan, Mahendragiri, and parts of Malkangiri and Phulbani districts. The high biological richness in these regions is a result of microclimatic conditions and conservation measures. High biodiversity zones are also found in some parts of Malkangiri, Koraput, Rayagada, Gajapati, Phulbani, and Nuapada districts. These areas fall in the transition of sal and non-sal zones. The medium biodiversity regions are also seen to be available in parts of Nabarangpur, Kalahandi, Phulbani, Nayagarh, Khordha, Angul, Bargarh, and Keonjhar districts. The low biodiversity zones are found in parts of Sambalpur, Deogarh, Sundergarh, Puri, and Keonjhar districts.

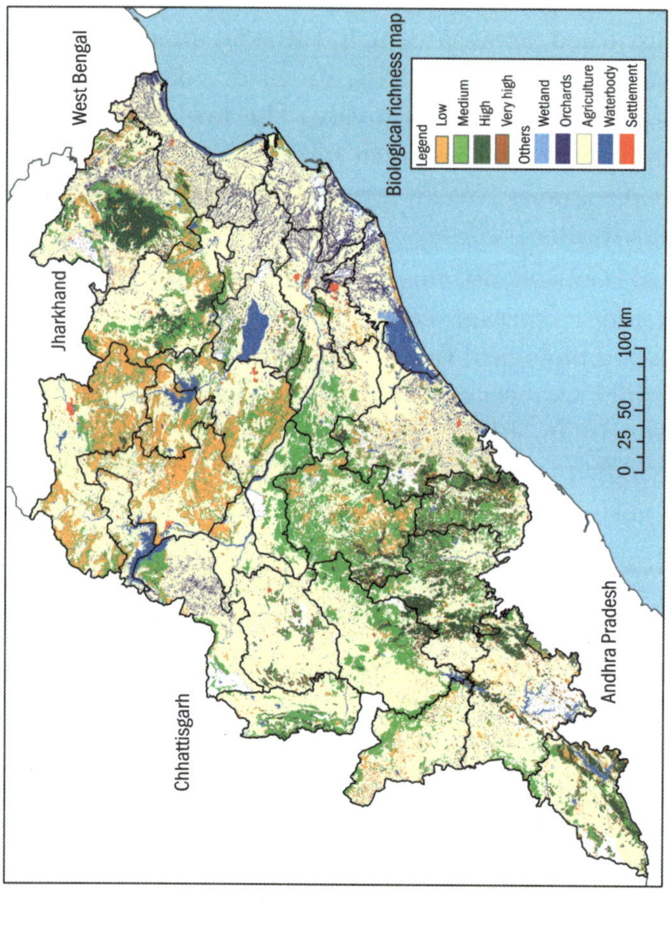

Figure 7 *Biological richness map of Odisha*

Medicinal Plant Richness Areas

Similipahar hill range

Similipal lies between 86°10' E to 86°20' E and 21°27' N to 22°01' N on the north-eastern part of the state and covers an area of 2750 km^2. There are 1012 wild taxa and 64 cultivated species spread over 168 families (Saxena and Brahmam 1989). The vegetation of Similipal forest can be classified into five types according to Champion and Seth (1968). Many species have been found to be new records for the flora of Odisha. These new recorded floras are: *Colysis pedunculata* (Hook. & Grev.) Ching, *Salomonia cantoniensis* Lour, *Oxalis latifolia* Kunth, *Ziziphus glabrata* Heyne ex Roth, *Cissus assamica* (Lawson) Craib, *Luffa cylindrical* (L.) Roem. var. *minor* Chakravarthy, *Anaphalis adnata* DC, *Blumea aromatica* (Wall.) DC, *Blumea clarkei* Hook.f., *Embelia floribunda* Wall., *Cynanchum tunicatum* (Retz.) Alston, *Toxocarpus kleinii* W. & A, *Cordia wallichii* G. Don, *Aneilema ovalifolium* (Wight) Hook.f. and C. B. Clarke, and *Rhapidophora glauca* schott.

Some species are found exclusively in this area and nowhere else in the state and include *Dicrocephala integrifolia, Hypericum gaitti, Cyathia spinulata, Gomphostemma parviflorum, Neocinnamomum caudatum, Saphora bekeri,* and *T. kleinii.*

Gandhamardan hill range

The Gandhamardan hill range lies between 82°46' E to 83°01' E and 20°48' N to 20°55' N in Bolangir and Baragarh districts bordering Chhattisgarh. It covers an area of nearly 300 km^2 with the hill ranging 250–1050 m in height. There are as many as 912 species of flowering plants present in this hill range and more than 500 medicinal plant species (Pattanaik and Reddy 2007; Reddy and Pattanaik 2009). Some interesting taxa of the Gandhamardan area are *Erythrina resupinata, Tylophora fasciculata, Heterostemma tanjorense, Actinodaphne angustifolia, Pittosporum wightii,* and *Protium serratum.*

Mahendragiri hill range

The Mahendragiri hills are an important part of the Eastern Ghats and lie in the Gajapati district. It is the second highest peak in Odisha at an elevation of 1501 m. Diverse flora is found in these hill ranges since the subtropical climate favours some temperate species. Some of the Lower Himalayan and South Indian species are also found here and include *Maoutia puya, Viola serpens, Viola betonicifolia, Clematis roylei, Pittosporum nepalensis, Anotis calycina, Ajuga maerosperma, Thalictrum foliolosum, Rubus ellipticus, Sarcococca trinervia, Zanthoxylum armatum, Ophiopogon intermedius, Tephrosia roxburghiana, Sophora interrupta, Litsea latea, Neolitsea zeylanica, Diospyros candolleana, Pavetta brevifolia,* and *Oldenlandia gamble.* Many new species have been recorded for the flora of Odisha and include *Aganosma cymosa, Cyanotis vaga, Justicia glabra, Moutiya puya, Paspalum conjugatum, Plectranthus nilhericus, Psychotria fulva, Neanotis quadrilocularis, Ophiorhiza trichocarpa,* and *Eulalia quadrinervis.* Three new species were reported from this hill during 1970–80. These plant species are *Dimeria mahendragiriensis* Ravi, H. O. Saxena & Brahmam, *Habenaria panigrahiana* S. Misra, and *Habenaria panigrahiana* var. *parviloba* S. Misra. The hills look lush and green in every season because they contain several types of forests. The floristic composition of this hill is unique and found nowhere else in Odisha.

Deomali hill range

The Deomali hill range has the highest peak in the Eastern Ghats and lies in Koraput district. It is the tallest hill in the Eastern Ghats as well as in Odisha, with a height of 1672 m. The vegetation can be divided into sub-tropical broad leaved hill forest, moist deciduous forest, semi-evergreen forest, and shrub savannah. Some species specific to this hill and not found elsewhere in the state are *Habenaria grandifloriformis, Emilia zeylanica,* and *Gynura lycopersicifolia.* Many tribals live here and exploit the plant resources for their food, household, and medicine.

Saptasajya and Kapilasa hill ranges

Saptasajya and Kapilasa hill ranges are found in the Dhenkanal district of Odisha. The Saptasajya hill range is a constituent of seven hills and has a luxuriant forest cover. Most of the species composition are moist in nature as there are a number of rivulets in the hills. A plant species *Tetracera sarmentosa* (L.) Vahl subsp. *sumatrana* (Hoogl.) Hoogl. has been added to the flora of India as a new record from this hill range (Reddy and Pattanaik 2009). The Kapilasa hill range is situated on the other side of the Saptasajya. A lot of medicinal plants have been collected by local herbal healers from here in the recent past.

Kondakamberu hill range

The Kondakamberu hill range is found in the Malkangiri district of Odisha. This particular hill is still unexplored by botanists. During recent visits to the hill range, a number of medicinal plants were found. However, more field trips are required to explore the region. Nearly 70% of the medicinal plant resources are found in the wild. Most plant resources are confined to semi-evergreen, moist deciduous, and dry deciduous forests. Valley and riverine systems have more medicinal plants due to their moisture content. Some parts of the hills are also used for shifting cultivation, herbal medicine collection, and grazing by livestock from nearby villages. Such areas are extremely important from the conservation point of view and can be regarded as highly sensitive landscape units. Further studies in medicinal plant-rich areas are urgently required to focus on the ecology and patterns of biodiversity in such areas using high-resolution satellite data with a separate set of environmental variables. Key parameters such as the relative area under various landscape units and vegetation classes, species richness, and high diversity medicinal plant areas will need regular monitoring, for which baseline data has been generated through this study.

ACKNOWLEDGEMENTS

The authors are thankful to the Director, National Remote Sensing Centre, Hyderabad, and the Director, SACON, for their constant support and encouragement. The authors are also thankful to the Department of Biotechnology and the Department of Space, Government of India, for the financial assistance under which this particular biodiversity characterization study at landscape level was carried out in the Eastern Ghats.

REFERENCES

Abbott, I. 1980. Theories dealing with land birds on islands. *Advances in Ecological Research* 11:329–371

Ambastha, S. P. 1992. *The Useful Plants of India*. New Delhi: Publication and Information Directorate, Council of Scientific Industrial Research

Ana, S., J. Calestous, A. M. Carrie, H. J. Daniel, A. G. Michael, G. Rodrigo, A. L. Sarah, and V. R. Walter. 1993. *Biodiversity Prospecting: Using Genetic Resources for Sustainable Development*. Washington, DC: World Resources Institute

Champion, H. G., and S. K. Seth. 1968. *Revised Survey of Forest Types in India*. New Delhi: Government of India Press Publication

Farnsworth, N. R. 1988. Screening plants for new medicines. In *Biodiversity*, edited by E. O. Wilson and F. M. Peters, pp. 83–97. New York: Academic Press

FRLHT (Foundation for Revitalisation of Local Health Traditions) 1998. *Red Listed Medicinal Plants of South India*. Bengaluru: FRLHT

FSI (Forest Survey of India) 2003. *State of Forest Report 2003*. New Delhi: FSI, Ministry of Environment and Forests, Government of India

Gaston, K. J., T. M. Blackburn, and K. K. Goldewijk. 2003. Habitat conservation and global avian biodiversity loss. *Proceedings of Biological Sciences London* B 270:1293–1300

Heywood, V. H. (ed). 1995. *Global Biodiversity Assessment*. New York: Cambridge University Press

Johri, B. M., U. Singh, and A. M. Wadhwani. 1996. *Dictionary of Economic Plants in India*. New Delhi: Indian Council of Agricultural Research

McKinney, M. L. 1997. How do rare species avoid extinction? A paleontological view. In *The Biology of Rarity: Causes and Consequences of Rare-common Differences*, edited by W. E. Kunin and K. J. Gaston, pp.110–129. London: Chapman & Hall

Myers, N. 1993. Biodiversity and the precautionary principles. *Ambio* 22(2–3):74–79

Myers, N., R. A. Mittermeier, C. G. Mittermeier, G. A. B. Da Fonseca, and J. Kent. 2000. Biodiversity hotspots for conservation priorities. *Nature* 403:853–858

Nayar, M. P. 1989. In-situ conservation of wild flora resources. *Bulletin Botanical Survey of India* 29:319–333

Nayar, M. P. 1996. *'Hotspots' of Endemic Plants of India, Nepal and Bhutan*. Thiruvananthapuram: Tropical Botanic Garden and Research Institute

Pattanaik, C., and C. S. Reddy. 2007. Medicinal plant resources of Gandhamardan hill range, Orissa: an urgent need for conservation. *National Academy Science Letters* 30(1&2): 35–38

Purvis, A., J. L. Gittleman, G. Cowlishaw, and G. M. Mace. 2000. Predicting extinction risk in declining species. *Proceedings of the Royal Society of London* B 267:1947–1952

Reddy, C. S., and C. Pattanaik. 2009. An assessment of floristic diversity of Gandhamardan hill range, Orissa, India. *Bangladesh Journal of Plant Taxonomy* 16(1):29–36

Reddy, C. S., M. Brahmam, and V. S. Raju. 2006. Conservation prioritization of endemic plants of Eastern Ghats, India. *Journal of Economic and Taxonomic Botany* 30(4):755–772

Reddy, C. S., K. N. Reddy, and S. N. Jadhav. 2001. *Threatened Medicinal Plants of Andhra Pradesh, India*. Hyderabad: Environment Protection Training and Research Institute

Rodrigues, A. S. L., S. J. Andelman, M. I. Baker, L. Boitani, T. M. Brooks, R. M. Cowling, L. D. C. Fishpool, G. A. B. daFonseca, K. J. Gaston, M. Hoffman, J. S. Long, P. A. Marquet, J. D. Pilgrim, R. L. Pressey, J. Schipper, W. Sechrest, S. N. Stuart, L. G. Underhill, R. W. Waller, M. E. J. Watts, and X. Yan. 2004. Effectiveness of the global protected area network in representing species diversity. *Nature* 428:640–643

Roy, P. S., M. S. R. Murthy, S. Singh, C. S. Jha, M. C. Porwal, S. Ashutosh, S. P. S. Khuhwah, C. S. Reddy, G. S. Pujar, S. Trivedi, A. Roy, H. Karnataka, M. M. Kimothi, M. K. Behera, P. K. Joshi, G. Rajashekar, and R. Paliwal. 2008. *Biodiversity Characterization at Landscape Level using Satellite Remote Sensing and Geographic Information System*. Hyderabad: National Remote Sensing Centre, Indian Space Research Organization, Department of Space, Government of India. [DOS-DBT User Manual]

Saxena, H. O. and M. Brahmam. 1983. Rare and endemic flowering plants of Orissa. In *An Assessment of Threatened Plants of India*, edited by S. K. Jain and R. R. Rao. Kolkata: Botanical Survey of India

Saxena, H. O., and M. Brahmam. 1989. *The Flora of Similipahar, Orissa*. Bhubaneswar: Regional Research Laboratory

Saxena, H. O., and M. Brahmam. 1996. *The Flora of Orissa, vol.* 1–4. Bhubaneswar: Odisha Forest Development Corporation Ltd

Thomas, C. D., A. Cameron, R. E. Green, M. Bakkenes, L. J. Beaumont, Y. C. Collingham, B. F. N. Erasmus, M. Ferreira deSiqueira, A. Grainger, L. Hannah, L. Hughes, B. Huntley, A. S. V. Jaarsveld, G. F. Midgley, L. Miles, M. A. Ortega-Huerta, A. T. Peterson, O. L. Phillips, and S. E. Williams. 2004. Extinction risk from climate change. *Nature* 427:145–148

UNEP-WCMC (United Nations Environment Programme-World Conservation Monitoring Centre). 2000. *Global Biodiversity: Earth's Living Resources in the 21st Century*. Cambridge: World Conservation Press

CHAPTER 6

Chemical Impact of *Parthenium hysterophorus* Linn. on Some Economically Important Vegetables

Zubaida Yousaf and Zeenat Haq

INTRODUCTION

Parthenium hysterophorus of the Asteraceae family is known by many common names such as Congress grass, Carrot grass (Oudhia 1998), White top, Santa Maria, Fever few, and rag weed. It is native to Mexico, and Central and South America (McFadyen 1992). It is an annual (under certain conditions a short-lived perennial), erect, ephemeral herb with vigorous growth (McFadyen 1992). The juvenile stage exists as a rosette with large dark green leaves and reduced stem, leaves fully covered with trichomes (hairs), and capitulums with five fertile ray florets. Highly prolific, an average plant can produce 15 000 seeds and large plants are known to produce 100 000 seeds (McFadyen 1992). More than 340 million *Parthenium* weed seeds per hectare can be present in the surface soil (CRC Weed Management 2003).

Parthenium, a highly invasive exotic species, has become an extremely serious agricultural and rangeland weed because it aggressively colonizes disturbed sites and invades open spaces. It has a major impact on pasture and cultivated lands, and is a threat to indigenous biodiversity. This plant is competing with some of the alien species such as *Cannabis sativa*, which does not grow in the vicinity of *Parthenium* (Hussain and Khatoon 2000).

Adverse effects of *Parthenium* have been documented on humans as well as on animals. It is known to be allergenic to some people (Shabbir and Bajwa 2004) and its consumption by livestock can taint meat. It is known to cause asthma, bronchitis, allergic contact dermatitis, and hayfever in humans, and skin rashes in animals. It is responsible for bitter milk disease in cattle, buffaloes, and goats that are fed on grass mixed with *Parthenium*. Growth suppression effect of this weed was found in some vegetable crops also (Shabbir and Bajwa 2004).

The allelopathic properties of *Parthenium* weed have been well demonstrated. These are largely attributed to the presence of parthenin, a sesquiterpene lactone of pseudguanolide nature found in various parts of *P. hysterophorus* (Kanchan and Jayachandra 1980; Kohli, Rani, and Verma 1993). *P. hysterophorus* is an allelopathic weed that reduces the growth and productivity of associated vegetation due to the release and accumulation of phytotoxins from its decomposing biomass and root exudates in the soil during crop growth (Narwal, Palaniraj, Sati, *et al.* 2003). The current research has been designed to evaluate the effect of aqueous extracts of *P. hysterophorus* on the germination and growth rates of *Lactuca sativa, Spinacea oleracea, Brassica rapa*, and *Pisum sativum*.

MATERIALS AND METHODOLOGY

Plant Material Collection

The plant material (*P. hysterophorus)* was collected from a field area of the Pir Mehr Ali Shah Arid Agriculture University, Rawalpindi in Pakistan. The allelopathic effect of cold water extracts from fresh leaves, roots, stems, flowers, and the whole plant was tested on the germination and seedling growth of *L. sativa, S. oleracea, B. rapa*, and *P. sativum* in the university's Department of Botany laboratory.

Preparation of Aqueous Extracts

Plant material was collected and washed thoroughly under running water to remove dust and other undesired organic material. The material was crushed with a mortar and pestle and soaked. For the whole plant treatment, equal weights of crushed leaves, roots, stems, and flowers were mixed thoroughly for homogeneity and soaked in water. Extracts were filtered using Whatman filter paper No.1. The filtrates were stored in a refrigerator at 8°C for further use. The effect of aqueous extracts was tested against the germination and seedling growth of *L. sativa*, *S. oleracea*, *B. rapa*, and *P. sativum* using standard filter paper bioassay techniques.

Sowing of Seeds

Sterilized Petri dishes were lined with two layers of Whatman filter paper No.1 to form the seed bed. The paper was soaked with 3 ml of plant extracts and healthy, sterilized seeds of *L. sativa*, *S. oleracea*, *B. rapa*, and *P. sativum* were spread uniformly on the filter paper in individual Petri dishes.

STATISTICAL ANALYSIS

The data was analysed using Statistical Procedure for Social Sciences (SPSS) software, version 11, and Microsoft Excel. As the experimental design was Randomized Complete Split-block Design, single-factor ANOVA (Analysis of variance) was calculated to investigate the significance of alleopathic activity. The level of significance was 0.05. The percentage growth of the radicles and plumules of the test plants under the influence of the aqueous extract of various herbs is represented by the bar graph in Figure 1.

RESULTS

The allelopathic effect of *P. hysterophorus* was analysed on the basis of the growth rates of the radicle and hypocotyl of *L. sativa*, *S. oleracea*, *B. rapa*, and *P. sativum* seedlings in different extracts with varying concentrations of *P. hysterophorus*.

Effect of *Parthenium hysterophorus* on *Lactuca sativa*

Table 1 shows the growths of the radicles and hypocotyls of *L. sativa* seedlings in different extracts with varying concentrations of *P. hysterophorus*. The whole plant extracts of *P. hysterophorus*

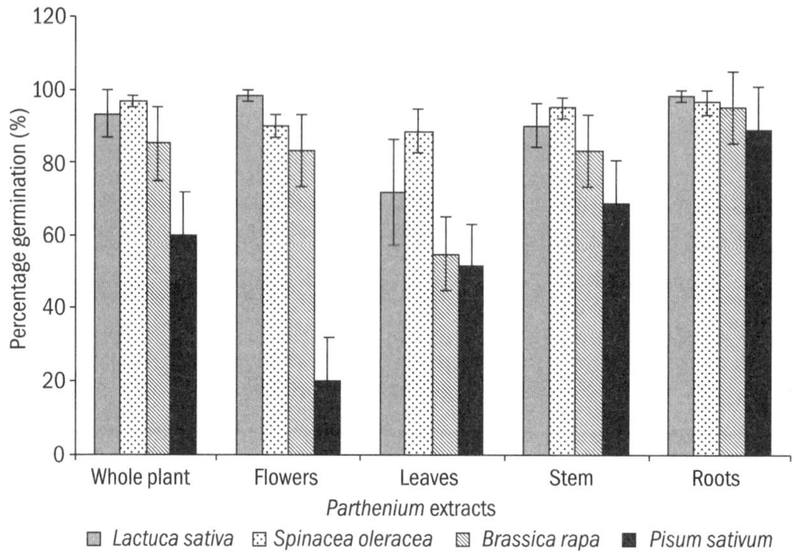

Figure 1 *Effect of aqueous extracts of* Parthenium hysterophorus *on seed germination of* Lactuca sativa, Spinacea oleracea, Brassica rapa, *and* Pisum sativum

showed variable stimulatory effects on the hypocotyl growth of *L. sativa*. At 10% concentration, significantly ($P < 0.05$) high growth rate (2.597 ± 0.09) was found, compared to control (1.347 ± 0.08). A similar stimulatory effect was also observed with 5% and 15% whole plant extracts but the least stimulation was noticed with 15% concentration. The stimulatory effect of the whole plant extract was also significantly high compared to the flower extract (5%) and the leaf extract (15%).

The flower extract exhibited a significant ($P < 0.05$) enhancing effect on the growth of the hypocotyls of *L. sativa* with respect to control in 5% and 15% concentrations. The effect of 10% flower extract is non-significant compared to control. It was observed that the leaf extract (5%) of *P. hysterophorus* significantly enhanced the hypocotyl growth of *L. sativa*. As the concentration of leaf extract was increased, a marked inhibition in hypocotyl growth was found. The 15% leaf extract strongly inhibited the hypocotyl growth (1.150 ± 0.04) compared to other extracts.

Table 1 *Growth rate of* Lactuca sativa *(lettuce) in varying concentrations of different parts of* Parthenium hysterophorus *L.*

Treatment	*Concentration (%)*	*Hypocotyl (Mean in cm ± SE)*	*Radicle (Mean in cm ± SE)*
Control	0	1.347 ± 0.08	2.060 ± 0.11
Whole plant extract	5	2.320 ± 0.13	2.040 ± 0.09
	10	2.597 ± 0.09	1.343 ± 0.06
	15	2.083 ± 0.14	1.240 ± 0.06
Flower extract	5	2.497 ± 0.10	1.650 ± 0.06
	10	1.380 ± 0.07	1.203 ± 0.04
	15	1.777 ± 0.07	1.147 ± 0.05
Leaf extract	5	2.337 ± 0.10	1.660 ± 0.10
	10	1.780 ± 0.07	1.047 ± 0.05
	15	1.150 ± 0.04	0.533 ± 0.04
Stem extract	5	1.820 ± 0.11	1.657 ± 0.09
	10	2.237 ± 0.11	1.570 ± 0.08
	15	2.060 ± 0.05	1.307 ± 0.07
Root extract	5	2.000 ± 0.09	2.527 ± 0.09
	10	2.237 ± 0.04	2.710 ± 0.11
	15	2.063 ± 0.07	2.113 ± 0.18

Note Significant level (α value) = 0.05

The growth rate of the hypocotyl was significantly (P < 0.05) different from control at all concentrations of *P. hysterophorus* stem extracts. The maximum hypocotyl growth was found in 10% stem extract treatment. The minimum stimulatory effect was observed in 5% stem extract. The effect of stem extract at all concentrations was non-significantly different (P > 0.05) from other extracts. It appeared that the stem extract showed a significant stimulatory effect compared to the leaf (15%) and flower (10%) extracts of *P. hysterophorus*. The hypocotyl of *L. sativa* showed a significantly high (P < 0.05) promotive effect on the growth rate in the root extract of *P. hysterophorus*. The effect was significant (P < 0.05) in contrast to control, leaf extract (15%), and flower extract (10%). However, non-significant difference was observed with respect to all other aqueous extracts. The stimulatory effect was more pronounced in 10% root extract.

Radicle growth in *L. sativa* in 5% whole plant extract exhibited non-significant (P > 0.05) inhibitory effect, while in 10% and 15% whole plant extract, it showed significant inhibition compared to control. The maximum reduction was found in 15% concentration. Significant inhibitory effect was seen with 10% and 15% concentrations of whole plant extract in comparison to root extract. The effect of *P. hysterophorus* flower extracts was found inhibitory to the growth rate of *L. sativa*. This inhibitory effect was more pronounced with increasing concentrations. The maximum inhibition was observed in 15% aqueous flower extract. However, all concentrations showed significant ($P < 0.05$) reduction of radicle growth with respect to control, root extracts, and 5% whole plant extract. It was observed that the inhibitory effect of flower extracts was significantly ($P < 0.05$) less than that of 15% leaf extract.

Radicle growth was strongly inhibited by the leaf extracts of *P. hysterophorus*. The effect was more profound when concentration was increased. Although the 5% and 10% leaf extracts were non-significantly different from each other, they significantly inhibited the radicle growth of *L. sativa* in comparison to control. The 15% leaf extract showed significant ($P < 0.05$) inhibition compared to the other two concentrations (5% and 10%), control, and extracts of all other parts of *P. hysterophorus*. The effects of 5% and 10% leaf extracts showed pronounced inhibition in comparison to root extracts of *P. hysterophorus*. The stem extracts of *P. hysterophorus* showed significant ($P < 0.05$) inhibitory effects compared to control. The inhibitory effect of stem extracts was similar to that of the whole plant extract (10%, 15%), flower extract, and leaf extract (5%, 10%). This effect was less than that of 15% aqueous leaf extract but significantly greater than that of root extracts.

It was found that root extracts significantly stimulated the radicle growth of *L. sativa* compared to all other aqueous extracts of different parts of *P. hysterophorus*. Root extract (10%) showed significantly ($P < 0.05$) the maximum stimulatory effect on growth rate (2.710 ± 0.11) compared to control (2.060 ± 0.11) and all other aqueous extracts of *P. hysterophorus*. However, the seed germination of *L. sativa* was significantly ($P < 0.05$) inhibited by the leaf extracts of *P. hysterophorus* (71.67 ± 14.55). The whole plant extracts, flower extracts, stem extracts, and root extracts did not show any pronounced response to the germination of *L. sativa* seeds.

Effect of *Parthenium hysterophorus* on *Spinacea oleracea*

Table 2 shows the growth of the radicle and hypocotyl of *S. oleracea* (spinach) seedlings in different extracts with varying concentrations of *P. hysterophorus.* It was noticed that the hypocotyl growth of *S. oleracea* was promoted in the aqueous whole plant extract and saw the maximum stimulatory effect in 15% concentration (3.591 ± 0.10). The flower extracts caused reduction in hypocotyl growth and this reduction increased with increase in concentration, so that the least inhibition was observed in 5% flower extract and the most in 15% flower extract. The 15% flower extract showed the greatest inhibitory effect on growth rate (1.477 ± 0.08) compared to control (2.497 ± 0.16) and all other aqueous extracts of *P. hysterophorus*.

The growth rate of *S. oleracea* hypocotyl was suppressed in the leaf extract. Minimum suppression was found in 5% concentration while the maximum ($P < 0.05$) inhibition was observed in 15% concentration. Stem extracts and root extracts of *P. hysterophorus*

Table 2 *Growth rate of* Spinacea oleracea *(spinach) in varying concentrations of different parts of* Parthenium hysterophorus *L.*

Treatment	*Concentration (%)*	*Hypocotyl (Mean in cm ± SE)*	*Radicle* (Mean in cm ± SE)*
Control	0	2.497 ± 0.16	4.463 ± 0.48
Whole plant extract	5	2.977 ± 0.16	4.518 ± 0.44
	10	2.521 ± 0.15	2.480 ± 0.43
	15	3.591 ± 0.10	4.111 ± 0.27
Flower extract	5	1.911 ± 0.20	2.479 ± 0.29
	10	1.885 ± 0.24	1.489 ± 0.26
	15	1.477 ± 0.08	0.3913 ± 0.05
Leaf extract	5	2.099 ± 0.06	1.782 ± 0.18
	10	1.627 ± 0.16	0.6707 ± 0.07
	15	1.579 ± 0.10	0.3920 ± 0.04
Stem extract	5	2.089 ± 0.21	3.965 ± 0.40
	10	3.219 ± 0.19	4.039 ± 0.38
	15	2.464 ± 0.15	3.550 ± 0.31
Root extract	5	3.574 ± 0.22	5.675 ± 0.37
	10	2.730 ± 0.12	4.081 ± 0.28
	15	2.747 ± 0.18	4.785 ± 0.50

* The data for radicle has been transformed sqrt (v + 1)

showed non-significant ($P > 0.05$) stimulatory effect on hypocotyl growth. Similarly, root extracts (5%) of *P. hysterophorus* exhibited the maximum stimulatory effect on the growth rate (3.574 ± 0.22) compared to 15% whole plant extract (3.591 ± 0.10).

The whole plant extracts of *P. hysterophorus* showed variable results in all three concentrations. In the 5% whole plant extract, radicle growth exhibited non-significant ($P > 0.05$) promotion compared to control, whole plant (10%), stem extracts, and root extracts of *P. hysterophorus.* This positive impact is significantly different from extracts of the flower and leaf. The 10% whole plant extract showed a significant inhibitory effect on radicle growth compared to control, whole plant (5%, 15%), flower extract (15%), leaf extract (10%, 15%), stem extracts, and root extracts. This effect was somewhat similar to that of the flower extract (5%, 10%) and leaf extract (5%). The 15% concentration revealed the maximum non-significant ($P > 0.05$) stimulatory effect in comparison to control, whole plant (5%), stem extracts, and root extracts, but exhibited significant positive effect with respect to flower extracts and leaf extracts.

The inhibition caused by flower extracts was prominent. It was noticed that the inhibitory effect was the least in 5% concentration and greatest in 15% concentration. It showed that 5% and 10% flower extracts inhibited the growth of radicle significantly ($P < 0.05$) compared to control, whole plant extract (5%, 15%), stem extracts, and root extracts. This negative effect is significant in comparison to whole plant extract (10%) and leaf extract (5%) but significantly ($P < 0.05$) less than that of flower extract (15%) and leaf extract (10%, 15%). It was also observed that 15% flower extract showed the maximum inhibitory effect in radicle growth (0.3913 ± 0.05).

The growth rate of the radicle in *S. oleracea* was also inhibited in the leaf extracts of *P. hysterophorus.* A markedly significant ($P < 0.05$) inhibition in the radicle growth was found in 10% (0.6707 ± 0.07) and 15% (0.3920 ± 0.04) leaf extracts with respect to control and all other aqueous extracts of *P. hysterophorus.* The 5% leaf extract showed reduction in radicle growth but significantly less than that of 10% and 15% leaf extracts and in comparison to the whole plant (5%, 15%), stem, and root extracts. This impact of

5% leaf extract was non-significant (P > 0.05) with respect to whole plant extract (10%) and flower extract (5%, 10%).

The stem extracts showed non-significant (P > 0.05) enhancement in radicle growth compared to control, whole plant extract (5%, 15%), and root extracts. The results were significantly different from whole plant extract (10%), flower extracts, and leaf extracts. The positive impact of stem extracts on the radicle growth of *S. oleracea* was similar in all three concentrations but a non-significant (P > 0.05) greater stimulatory effect was observed in 10% stem extract. It appeared that the root extracts caused a significant (P < 0.05) stimulatory effect in comparison to whole plant extract (10%), flower extracts, and leaf extracts. The effects of different concentrations of root extracts were non-significantly (P > 0.05) different from each other as well as from the control, whole plant extract (5%, 15%), and stem extracts. The maximum stimulation of radicle growth (5.675 ± 0.37) was observed in 5% root extract and the least in 10% concentration.

Germination (88.33 ± 6.02) of *S. oleracea* seeds was inhibited in the leaf extracts of *P. hysterophorus.* However, whole plant extract, flower extracts, stem extracts, and root extracts did not show any significant effect on seed germination.

Effect of *Parthenium hysterophorus* on *Brassica rapa*

Table 3 shows the growths of the radicle and hypocotyl of *B. rapa* (turnip) seedlings in different extracts with varying concentrations of *P. hysterophorus.* The overall impinge of whole plant extracts was non-significantly (P > 0.05) negative compared to control, 5% flower extract, stem extracts (10%, 15%), and root extracts. This inhibitory effect was non-significantly similar to that of flower extract (10%, 15%), leaf extracts, and 5% stem extract. At the lowest concentration, 5% whole plant extract, the greatest suppression of hypocotyl growth (2.173 ± 0.18) was observed. This reduction in hypocotyl growth non-significantly (P > 0.05) decreased when the concentration was increased so that, at 15% concentration, the least inhibition was noticed.

Table 3 *Growth rate of* Brassica rapa *(turnip) in varying concen-trations of different parts of* Parthenium hysterophorus *L.*

Treatment	*Concentration (%)*	*Hypocotyl (Mean in cm ± SE)*	*Radicle (Mean in cm ± SE)*
Control	0	2.047 ± 0.10	3.117 ± 0.18
Whole plant extract	5	1.127 ± 0.11	2.453 ± 0.18
	10	1.553 ± 0.14	1.990 ± 0.12
	15	1.647 ± 0.14	2.673 ± 0.18
Flower extract	5	2.507 ± 0.23	2.777 ± 0.27
	10	1.663 ± 0.11	2.313 ± 0.13
	15	1.267 ± 0.08	0.9833 ± 0.13
Leaf extract	5	1.967 ± 0.11	2.340 ± 0.17
	10	1.683 ± 0.07	1.450 ± 0.07
	15	1.460 ± 0.09	1.170 ± 0.14
Stem extract	5	1.473 ± 0.07	2.450 ± 0.17
	10	2.157±0.14	3.427 ± 0.25
	15	2.277 ± 0.16	2.670 ± 0.29
Root extract	5	2.407 ± 0.15	5.153 ± 0.23
	10	2.173 ± 0.18	3.823 ± 0.23
	15	1.773 ± 0.19	3.677 ± 0.39

The flower extract showed non-significant ($P > 0.05$) maximum stimulatory effect on growth rate (2.507 ± 0.23) at 5% concentration and an inhibitory effect at 10% and 15% concentrations compared to control. It was also noticed that the flower extracts exhibited non-significant ($P > 0.05$) impact compared to all other extracts of *P. hysterophorus*.

A decrease in the hypocotyl growth of *B. rapa* was seen in the leaf extracts. The maximum non-significant ($P > 0.05$) inhibition was observed in 15% aqueous leaf extract. Overall, the inhibitory effect of leaf extracts was non-significantly ($P > 0.05$) different from the control, flower extract (5%), stem extracts (10%, 15%), and root extracts (5%, 10%). Non-significant ($P > 0.05$) inhibition was found in 5% stem extract similar to that of whole plant extracts, flower extracts (10%, 15%), leaf extracts (10%, 15%), and 15% root extract. The 10% and 15% stem extracts showed non-significant ($P > 0.05$) promotive effect on the hypocotyl growth compared to control.

The 5% (2.407 ± 0.15) and 10% (2.173 ± 0.18) root extracts enhanced the hypocotyl growth of *B. rapa* non-significantly (P > 0.05) compared to control and all other extracts. This enhancement in hypocotyl growth was comparatively less than that in 5% flower extract. At 15% root extract, a slight non-significant (P > 0.05) decrease in growth rate was observed with respect to control. This decrease was non-significantly less than that of whole plant extracts, flower extracts (10%, 15%), leaf extracts (10%, 15%), and 5% stem extract.

The radicle growth of *B. rapa* was non-significantly (P > 0.05) suppressed compared to control in whole plant extracts. The greatest reduction was found in 10% whole plant extract and the least in 15%. It was also observed that the inhibitory effect of whole plant extracts was significantly (P < 0.05) less than in flower extracts (15%) and leaf extracts (10%, 15%). However, the impact was significant (P < 0.05) in comparison to 5% root extract. It was also observed that with increasing concentration, the severity of inhibition also increased. At 15% flower extract, significant (P < 0.05) suppression was found compared to control and all other corresponding extracts of *P. hysterophorus*. This inhibition was non-significantly different from 10% and 15% leaf extracts. The 5% and 10% flower extracts showed non-significant reduction in comparison to control, whole plant extracts, leaf extracts (5%), and stem extracts (5%, 15%). This effect of 5% and 10% flower extracts was significant (P < 0.05) with respect to 5% root extract.

The leaf extracts of *P. hysterophorus* showed a reduction in the hypocotyl growth of *B. rapa*. The 10% and 15% leaf extracts showed significant (P < 0.05) inhibitory effect compared to control and all other extracts of *P. hysterophorus* except 15% flower extract. The 5% leaf extract was significantly different from the other two concentrations, 5% root extract and 15% flower extract, while it was non-significantly different from control, whole plant extracts, flower extracts (5%, 10%), stem extracts, and root extracts (10%, 15%). The enhancement in the growth rate of radicle was found in stem extracts. The stem extracts showed significant (P < 0.05) stimulation compared to flower extracts (15%) and leaf extracts (10%, 15%). This stimulatory effect was significantly less than that of 5% root extract. The 10% stem extract showed greater promotion of radicle growth but it was non-significant in the other two concentrations and control.

The growth rate of radicle of *B. rapa* was significantly ($P < 0.05$) promoted in the root extracts. The 5% root extract significantly enhanced the radicle growth (5.153 ± 0.23) compared to control (3.117±0.18) and all other corresponding extracts of *P. hysterophorus*. The stimulatory effect was greatest in 5% concentration, then in 10% concentration, and the least in 15% concentration of root extract. The impact of 10% and 15% root extracts was non-significantly different from control, whole plant extracts, flower extracts (5%, 10%), 5% leaf extract, and stem extracts. However, it showed significant stimulation in comparison to flower extracts (15%) and leaf extracts (10%, 15%).

The whole plant extracts, flower extracts, leaf extracts, and stem extracts of *P. hysterophorus* exhibited inhibition in the seed germination of *B. rapa*. The maximum significant inhibition in germination rate (55.00 ± 18.95) was found in leaf extracts.

Effect of *Parthenium hysterophorus* on *Pisum sativum*

Table 4 shows the growths of the radicle and hypocotyl of *P. sativum* (pea) seedlings in different extracts with varying concentrations of *P. hysterophorus*. The effect of whole plant extracts on *P. sativum* was non-significantly ($P > 0.05$) stimulatory compared to control and all other extracts. The positive impinge of whole plant extracts was non-significantly different among all the concentrations. It was observed that this promotion in the radicle growth caused by whole plant extracts was also non-significantly similar to that of flower extracts (5%), leaf extracts (5%, 10%), stem extracts, and root extracts.

The flower extracts of *P. hysterophorus* showed variable results. In 5% flower extract, the hypocotyl growth of *P. sativum* showed a slight non-significant ($P > 0.05$) enhancement; but in 10% and 15% extracts, it exhibited reduction in hypocotyl growth. Maximum inhibition in growth rate (0.783 ± 0.18) was found in 15% flower extract. Both stimulatory and inhibitory impacts of flower extracts

Table 4 *Growth rate of* Pisum sativum *(pea) in varying concen-trations of different parts of* Parthenium hysterophorus *L.*

Treatment	*Concentration (%)*	*Hypocotyl (Mean in cm ± SE)*	*Radicle (Mean in cm ± SE)*
Control	0	1.223 ± 0.10	6.523 ± 0.64
Whole plant extract	5	1.453 ± 0.08	7.377 ± 0.32
	10	1.500 ± 0.16	8.617 ± 0.66
	15	1.547 ± 0.12	7.680 ± 0.33
Flower extract	5	1.317 ± 0.13	5.593 ± 0.54
	10	0.933 ± 0.20	2.940 ± 0.60
	15	0.783 ± 0.18	1.693 ± 0.35
Leaf extract	5	1.373 ± 0.23	4.273 ± 0.66
	10	1.453 ± 0.08	4.687 ± 0.39
	15	0.703 ± 0.18	2.217 ± 0.49
Stem extract	5	1.340 ± 0.25	6.670 ± 1.12
	10	1.923 ± 0.27	4.613 ± 0.72
	15	2.290 ± 0.29	6.287 ± 0.72
Root extract	5	1.857 ± 0.19	7.717 ± 0.88
	10	1.403 ± 0.24	6.040 ± 0.98
	15	1.447 ± 0.28	5.247 ± 0.98

Note Radicle and hypocotyle lengths have been transformed sqrt (v + 1)

were non-significantly (P > 0.05) different from control and all other extracts of *P. hysterophorus.*

The 5% and 10% leaf extracts exhibited a non-significant (P > 0.05) stimulatory effect compared to control. The maximum inhibition in growth rate (0.703 ± 0.18) was found in 15% leaf extract, non-significant (P > 0.05) compared to control (1.223 ± 0.10) and all other extracts of *P. hysterophorus.* This reduction in hypocotyl growth was non-significantly similar to those in flower extracts (10%, 15%). It was also observed that the stimulatory effect caused by 5% and 10% leaf extracts was comparatively similar to whole plant extracts, 5% flower extract, 5% stem extract, and root extracts (10%, 15%). This impact was non-significantly less than in the stem extracts (10%, 15%) and root extracts (5%).

It was noticed that stem extracts caused a stimulatory impinge on the hypocotyl growth of *P. sativum.* This enhancement caused by stem extracts was non-significantly (P > 0.05) different from all other extracts of *P. hysterophorus.*The maximum stimulation on growth rate (2.290 ± 0.29) was found in 15% stem extract compared to control (1.223 ± 0.10) and all other extracts.

The root extracts also exhibited a non-significant stimulatory effect compared to control and all other extracts of *P. hysterophorus*. The greater non-significant promotion was observed in 5% root extract than in 15%, and the least enhancement was found in 10% root extract. It was observed that the overall impact of whole plant extracts on the radicle growth of *P. sativum* was stimulatory. The maximum stimulatory effect on growth rate (8.617 ± 0.66) was observed in 10% whole plant extract. The effect of whole plant extracts was non-significantly ($P > 0.05$) different from control, 5% flower extract, leaf extracts (5%, 10%), stem extracts, and root extracts. However, it was significantly ($P < 0.05$) different from flower extracts (10%, 15%) and leaf extract (15%).

The flower extracts inhibited the growth of the radicle in all the three concentrations. The effect increased with the increase in concentration, and maximum inhibition in radicle growth (1.693 ± 0.35) was found in 15% flower extract compared to control (6.523 ± 0.64). The 5% flower extract showed non-significant ($P > 0.05$) inhibitory effect compared to control, whole plant extracts, stem extracts (5%, 15%), and root extracts. In 10% and 15% flower extracts, it was observed that there was significant ($P < 0.05$) reduction in radicle growth. This reduction was also significantly different from all other extracts of *P. hysterophorus* except the 15% leaf extract.

A significant inhibitory effect was found in 15% leaf extract compared to control and all other extracts of *P. hysterophorus* except the flower extracts (10%, 15%). The suppression in the radicle growth of *P. sativum* was also observed in 10% and 15% leaf extracts. This negative impact was non-significantly different from control, whole plant extracts, 5% flower extract, stem extracts, and root extracts. The 5% stem extract exhibited slightly non-significant enhancement in the radicle growth compared to control. However, 10% and 15% stem extracts showed non-significant inhibitory effect compared to control. The maximum suppression was found in 10% stem extract. The effect of stem extracts was significantly different from the leaf extract (15%) and flower extracts (10%, 15%).

It was revealed that the root extract at 5% concentration showed a significantly stimulatory effect in comparison to flower extracts (10%, 15%) and leaf extract (15%). This impact was non-significantly different from the control, whole plant extracts, 5% flower extract, leaf extracts (5%, 10%), and stem extracts. In 10% and 15% root extracts, there seemed to be non-significant inhibitory effect with respect to control. This inhibitory effect was significantly less than the leaf extract (15%) and flower extracts (10%, 15%). The seed germination of *P. sativum* was affected significantly in almost all the extracts of *P. hysterophorus*. However, the maximum effect on seed germination (19.99 ± 16.79) was observed in the flower extracts.

DISCUSSION

The whole plant extract (10%) of *P. hysterophorus* showed a significant stimulatory effect ($P < 0.05$) on the hypocotyl growth (2.597 ± 0.09) of *L. sativa* in comparison to control (1.347 ± 0.08). However, strong inhibition in hypocotyl growth was only found in the leaf extracts with increasing concentration. The maximum inhibition was observed in 15% leaf extract (1.150 ± 0.07), which shows that the inhibitory effect is often dependent on concentration, as discovered by Iqbal, Furubayashi, and Fujii (2004). A similar inhibitory effect was found in the10% flower extract (1.380 ± 0.07). The whole plant extract, stem extracts, and root extracts at varying concentrations showed similarly stimulatory effects on the hypocotyl growth of *L. sativa* (Table 1). *L. sativa* exhibited maximum radicle growth rate in the root extract (10%) in comparison to control (2.710 ± 0.11 vs 2.060 ± 0.11). In leaf extract (15%), there was a significant strong inhibitory effect (0.533 ± 0.04) on the growth rate of the radicle. The whole plant, flower, and stem extracts of *P. hysterophorus* showed less inhibitory effect than the leaf extract, and the inhibition increased with increasing concentration (5% > 10% > 15%) (Table 1). Kohli, Kumari, and Saxena (1985) reported the same inhibitory effect in cabbage caused by the leaves and inflorescence of *P. hysterophorus*. It was observed that the germination of *L. sativa* seeds was inhibited by the leaf extracts of *P. hysterophorus* (71.67 ± 14.55) (Figure 1).

The hypocotyl growth of *S. oleracea* was severely suppressed in the 15% flower extract but the effect was statistically non-significant (P > 0.05) with respect to control (1.477 ± 0.08 vs 2.497 ± 0.16). The non-significant stimulatory effect was also found in the case of 15% whole plant extract (3.591 ± 0.10) and 5% root extract (3.574 ± 0.22) (Table 2). The stem, root, and whole plant extracts of *P. hysterophorus* exhibited non-significant stimulatory effect on the hypocotyl growth of *S. oleracea*. On the other hand, the flower and leaf extracts showed non-significant suppression. In the radicle growth of *S. oleracea*, it was noticed that 15% flower extract (0.391 ± 0.05), 10% leaf extract (0.670 ± 0.07), and 15% leaf extract (0.392 ± 0.04) exhibited significant (P < 0.05) maximum inhibition. Dongre, Singh, and Chaube (2004) also reported that 10% leaf extract caused a reduction in the seedling growth of *Phaseolus mungo* (black gram). The maximum non-significant enhancement in the growth rate of radicles was found in 5% root extract in comparison to control (5.675 ± 0.37 vs 4.463 ± 0.48) (Table 2). The seed germination (88.33 ± 6.02) of *S. oleracea* was slightly suppressed in the leaf extracts compared to the other extracts of *P. hysterophorus* (Figure 1). Gupta (1996) observed a similar reduction in the radicle and hypocotyl growth of spinach (*Amaranthus gangeticus*) caused by foliar leachates of *P. hysterophorus*.

The growth rate of the hypocotyl of *B. rapa* was non-significantly (P > 0.05) restricted in the 5% whole plant extracts compared to control (1.127 ± 0.11 vs 2.047 ± 0.10). It was also demonstrated that the maximum non-significant growth of the hypocotyl was found in 5% flower extract (2.507 ± 0.23) (Table 3). The whole plant extracts and leaf extracts showed non-significant inhibitory effect. The flower extracts also exhibited inhibitory effect but at higher concentrations. Bajwa, Shafique, and Javaid (2004) reported that flower extracts of 10%, 30%, and 50% inhibited the seedling growth of the sunflower. Stem extracts exhibited non-significant stimulatory effect at higher concentration, while root extracts showed a similar inhibitory effect at lower concentration. The radicle growth was affected significantly (P < 0.05) in 15% flower extract (0.9833 ± 0.13), 10% leaf extract (1.45 ± 0.07), and 15% leaf extract (1.170 ± 0.14) compared to control. The effect caused by leaf extracts were supported by

Prasad (2004) who reported that leaf extracts show deleterious effects on plant height and days of flowering of *B. rapa*. Significant stimulation in the radicle growth was observed in *B. rapa* in 5% root extract compared to control (5.153 ± 0.23 vs 3.117 ± 0.18) (Table 3). Radicle growth was restricted in the whole plant extracts, flower extracts, and leaf extracts. It has been observed that the stem extracts showed both inhibitory and stimulatory effects at different concentrations. The aqueous root extracts exhibited the maximum stimulation in the radicle growth of *B. rapa*. It was noticed that the seed germination of *B. rapa* was significantly inhibited in the leaf extracts of *P. hysterophorus* (55.00 ± 18.95) (Figure 1).

It was noticed in *P. sativum* that the maximum promotion in the hypocotyl growth was found in 15% stem extract compared to control (2.290 ± 0.29 vs 1.223 ± 0.10). The non-significant ($P > 0.05$) reduction in the growth rate of hypocotyl was found in the 10% flower extract (0.933 ± 0.20), 15% flower extract (0.783 ± 0.18), and 15% leaf extract (0.703 ± 0.18) in comparison to control (Table 4). The whole plant extracts, stem extracts, and root extracts generally showed non-significant promotion in hypocotyl growth. It has been noticed that radicle growth showed the maximum non-significant stimulatory effect in 10% whole plant extract (8.617 ± 0.66 vs 6.523 ± 0.64) (Table 4). In the 10% flower extract (2.940 ± 0.60), 15% flower extract (1.693 ± 0.35), and 15% leaf extract (2.217 ± 0.49), a significant ($P < 0.05$) suppression of radicle growth was observed. The overall impact of whole plant extracts was stimulatory. The stem and root extracts also showed a stimulatory effect but at lower concentrations.

As the concentration increased, the inhibitory effects were observed. The flower and leaf extracts showed marked inhibitory effects. The flower extracts showed the maximum inhibition in the seed germination of *P. sativum* (19.99 ± 16.79) (Figure 1). Similar effects have been reported by Srivastava, Shukla, and Srivastava (1985) on barley, wheat, and peas. According to them, leaf and flower extracts inhibited germination and seedling growth but stem and root extracts were less inhibitory. Dhawan, Dhawan, and Dhawan (1995) also reported that the foliar extracts of *P. hysterophorus* completely inhibited the seed germination of okra, *Trifolium alexandrinum*, *Triticum aestivum*, and *Capsicum annum.*

CONCLUSION

P. hysterophorus is an exotic species widely distributed in Pakistan. It has a major impact on pasture and cultivated lands, and imposes a serious threat to native biodiversity due to its allelopathic effects. The present study concludes that the extract of the whole plant, leaves, flowers, stems, and roots of *P. hysterophorus* have mixed effects on the growth of four selected vegetables: *L. sativa, S. oleracea, B. rapa,* and *P. sativum.* Generally, the growth of these selected vegetables was inhabited in various concentrations of *P. hysterophorus* extract. In order to save the native biodiversity and reduce economic loss, attempts should be made to check the growth and further spread of *P. hysterophorus.*

REFERENCES

Bajwa, R., S. Shafique, and A. Javaid. 2004. Effect of foliar spray of aqueous extract of *Parthenium hysterophorus* on growth of sunflower. *International Journal of Agriculture and Biology* 6(3):474–478

CRC Weed Management. 2003. *Weed of the Month*: Parthenium hysterophorus. Australia: CRC for Australian Weed Management

Dongre, P. N., A. K. Singh, and K. S. Chaube. 2004. Allelopathic effects of weed leaf leachates on seed germination of black gram (*Phaseolus mungo L.*). *Allelopathy Journal* 14(1):65–70

Dhawan, P., S. R. Dhawan, and P. Dhawan. 1995. Allelochemic effect of *Parthenium hysterophorus* Linn. on the germination behaviour of some agriculture crops. *Flora and Fauna* 1:59–60

Gupta, A. 1996. Allelopathic influence of *Parthenium hysterophorus* L. over *Amaranthus gangeticus*. *Flora and Fauna* 2:57–58

Hussain, S. S., and S. Khatoon. 2000. Alien invasive species of Pakistan: list and brief information on preliminarily recognized invasive plant species in Pakistan. [Report of Workshop on Invasive Species in Pakistan, 17 September 1999, NARC, Islamabad, PARC, Pakistan]

Iqbal, Z., A. Furubayashi, and Y. Fujii. 2004. Allelopathic effect of leaf debris, leaf aqueous extract and rhizosphere soil of *Ophiopogon japonicus* Ker-Gawler on the growth of plants. *Weed Biology and Management* 4:43–48

Kanchan, S. D., and T. Jayachandra. 1980. Allelopathic effects of *Parthenium hysterophorus* L. II: leaching of inhibitors from aerial vegetative parts. *Plant and Soil* 55:61–66

Kohli, R. K., A. Kumari, and D. B. Saxena. 1985. Auto- and teletoxicity of *Parthenium hysterophorus* L. *Acta Universitatis Agriculturae Brno* 33:253–263

Kohli, R. K., D. Rani, and R. C. Verma. 1993. A mathematical model to predict tissue response to parthenin—an allelochemical. *Biology of Plant* 35:567–576

McFadyen, R. E. 1992. Biological control against parthenium weed in Australia. *Crop Protection* 11:76–88

Narwal, S. S., R. Palaniraj, S. C. Sati, H. S. Kadian, and D. S. Dahiya. 2003. Allelopathic plants: 8. *Parthenium hysterophorus* Linn. *Allelopathy Journal* 11(2):151–170

Oudhia, P. 1998. *Parthenium*: a curse for the bio-diversity of Chhattisgarh plains. In *Abstract of National Research Seminar on Bio-chemical Changes: An Impact on Environment* dated 30–31 July, p. 26, Mandla, India

Prasad, C. 2004. Influence of leaf extracts on cultivated turnip: VIII. Plant height, branches/plant and days to flower. *Cruciferae Newsletter* 25:53–54

Shabbir, A., and R. Bajwa. 2004. Distribution of *Parthenium* weeds (*Parthenium hysterophorus*) I: an alien invasive weed species threatening the biodiversity of Islamabad. *Weed Biology and Management* 2:133–142

Srivastava, J. N., J. P. Shukla, and R. C. Srivastava. 1985. Effect of *Parthenium hysterophorus* Linn. extract on the seed germination and seedling growth of barley and wheat. *Acta Botanica Indica* 13:194–197

CHAPTER 7

Genetic Factors Contributing to Plants Vulnerability and Extinction

Rizwana Khanum and Abdual Samad Mumtaz

INTRODUCTION

It is well known that both extrinsic and intrinsic characteristics have a significant impact on species to cope with evolutionary and environmental changes (Gray 1996; Huenneke 1991). This paper is an attempt to provide a summary of all the genetic factors that contribute to species extinction or vulnerability. The size of the population is the first clear indicator of the status of a species. Habitat fragmentation, especially human induced, is causing destruction in most parts of the world. Thus, species surviving in low populations and in isolated, fragmented habitats should be a priority for conservation (Frankham 1996).

Genetic factors have substantially and directly affected the vulnerability of a species because of inbreeding depression, which reduces the evolutionary potential of a species (Lynch 1996). It is also observed that inbreeding depression has the most deleterious effect on outbreeding species (Brook, Tonkyn, O'Grady, *et al.* 2002). As the effects of outbreeding depression are clearly less significant than inbreeding (Brook, Tonkyn, O'Grady, *et al.* 2002), they are not mentioned in detail here.

The comparison of vulnerable species (rare, endemic, threatened, extinct) as reviewed previously (Dodd and Helenurm 2002; Qu, Li, Luo, *et al.* 2007) shows indirectly the importance of all the mentioned factors for the rarity of a particular species. The implications for conservation are critically reviewed and it is found that the genetic analysis of a particular species is mandatory for the management of a vulnerable species.

GENETIC CHARACTERS AND EXTINCTION

Effect of Inbreeding Depression

Inbreeding and fitness characters

Inbreeding has long been known to reduce reproduction and survival in naturally outbreeding species (inbreeding depression). Inbreeding has serious effects on all aspects of reproduction and survival, including sperm production, mating ability, female fecundity, juvenile survival, mothering ability, age at sexual maturity, adult survival in animals, and in related components in plants (Brook, Tonkyn, O'Grady, *et al.* 2002).

The majority (about 90%) of the results show that inbred specimens have weaker reproductive fitness characters than non-inbred ones. Fitness comparisons of controlled self- and outcross-pollination in plants from a single population showed the existence of outbreeding depression or reduced fitness of progeny in outcrossed compared to the selfed ones (Quilichini, Thompson, and Debussche 2001). Generally, poor fertility between individuals of distant populations has been attributed to "outbreeding depression" caused by the breakdown of co-adapted gene complexes or unfavourable epistatic relationships (Lynch 1991; Fontdevila and Moya 1999; Fenster and Galloway 2000).

Quilichini, Thompson, and Debussche (2001) reported the presence of inbreeding depression—lower fitness of inbred progeny—and its significance for managing conservation programmes. The reason behind this is the localized adaptation or eco-adaptation of different genes, which become dislocated when outcrossing occurs; for example, in the plant species *Anchusa crispa.*

The species that inbred naturally suffer less reduction in all the fitness traits due to inbreeding depression (Frankham 2005). Similarly, inbreeding is estimated to erode heterozygosity more slowly in autotetraploids than in diploids (Glendinning 1989) and, under partial dominance, less inbreeding depression is expected due to the existence of partial heterozygotes (Husband and Schemske 1997) while inbreeding is avoided in all dioecious plant species. However, gene exchange between related plants is unavoidable within a population (Ellstrand and Elam 1993; Gaston and Kunin 1997; Tansley and Brown 2001).

Inbreeding and extinction

As inbreeding reduces fitness in terms of reproduction and survival, it is expected to increase extinction risks under situations where other factors are controlled or excluded. Inbreeding and loss of genetic diversity have been shown to contribute to extinction risk in the plant *Clarkia pulchella* (Newman and Pilson 1997). In the plant experiment, there was a large difference in extinction for a small difference in breeding co-efficient, so that the extinction rate for less inbred populations (F = 4%) was 25% while it was 69% for more inbred populations (F = 8%–9%) (Frankham 2005). In a controlled study of the butterfly *Glanville fritillary*, 42 populations were genotypically studied and it was observed that when all other extinction factors were removed or controlled, inbreeding was still an important indicator for extinction risk. Overall, inbreeding represents 26% of the extinction of the taxa (Saccheri, Kuussaari, Kankare, *et al.* 1998). Comparison of 20 threatened species for models with and without inbreeding for extinction using computer projections (Brook, Tonkyn, O'Grady, *et al.* 2002) showed greater extinction due to inbreeding depression in the taxa (Frankham 2005).

From a conservation perspective, small populations are particularly vulnerable to extinction due to low genetic variation, inbreeding depression, and stochastic events amongst others (Primack 1993). Changes such as habitat loss and reduction in population size can enhance the risk of extinction due to increase in inbreeding depression (Barrett and Kohn 1991; Young, Boyle, and Brown 1996). Similarly, the loss of genetic variation possibly due to genetic inbreeding and drift results in the extinction of the Seven Son flower (Hartl and Clark 1989). Apart from the risks due to inbreeding, fragmentation and decrease in size also enhanced this effect (Liu, Yang, Hao, *et al.* 2007; Zietsman, Dreyer, and Vuuren 2009).

Chang, Kim, Park, *et al.* (2004) studied endangered species and found that the effect of habitat fragmentation did not quite affect genetic variation, population size and results in extinction, and probably the factors that lie behind this are the high level of inbreeding along with high rate of seed production and selfing. The inbreeding coefficients (F) for six of the polymorphic loci were positive. Similar results were observed in threatened species (Hartl and Clark 1997; Hedrick 2000; Gapare, Aitken, and Ritland 2005; Eckstein, Oneill, Danihelka, *et al.* 2006; Chen, Li, Yin, *et al.* 2008). Obviously, various field studies are required to elaborate on these phenomena followed by an effective management strategy.

Rate of extinction and genetic factors

There are several studies to show that higher plant species that are endemic, rare, or threatened face the loss of genetic diversity as well as genetic variation (He, Rao, You, *et al.* 2000; Li, Xu, and He 2002; Cao, Yao, Ding, *et al.* 2006; Astorga, Vovides, Toledo, *et al.* 2009; Chen, Wang, Chen, *et al.* 2009). Similar observations were obtained for the level of genetic diversity in endangered pteridophytes (Hsu, Moore, and Chiang 2000; Landergott, Holderegger, Kozlowski, *et al.* 2001; Chen, Wang, Liu, *et al.* 2004; Chen, Liu, Wang, *et al.* 2005). Thus, genetic factors impact at a greater rate than expected. Among these, the impact of inbreeding on plant species is severe and a major cause of their extinction.

Lastly, small surviving populations that showed normal fitness traits, such as Chatham Island black robins (*Petroica traversi*), golden hamsters (*Mesocricetus auratus*), and Mauritius kestrels (*Falco punctatus)* survived bottlenecks of a single pair (Craig 1994; Groombridge, Jones, Bruford, *et al.* 2000; Elgar and Clode 2001; Frankham, Ballou, and Briscoe 2002). The repeated and continuous range of contraction of these species may experience severe bottlenecks and a subsequent genetic drift, which results in a series of problems such as fragmentation, isolation, and loss in genetic variation, all inducing extinction (Sherwin and Moritz 2000; Wang, Li, Hao, *et al.* 2004). Thus, inbreeding depression is more deleterious for outbreeding taxa.

Genetic Diversity: Loss and Evolution

The risk of extinction is enhanced with the loss of genetic diversity, which severely affects the ability of a population to evolve and cope with environmental changes. Environmental changes or evolutionary potentials are observed in almost all the species due to diverse reasons such as climatic change, pests, or parasites (Frankham and Kingsolver 2004). Changes in evolution and the history of different species have been documented by many authors (Frankel and Soule 1981; Lande 1995; Lynch 1996; Newman and Pilson 1997; Soule and Orians 2001; Knapen, Knaepkens, Bervoets, *et al.* 2003; Quilichini, Debussche, and Thompson 2004; Kadereit and Kadereit 2005).

According to Lynch and Lande (1998), the effective size of genetically secure population of an organism suggested from theoretical work would be 1000–5000 to retain maximum evolutionary potential and minimize the mutations. However, some recent work suggests that it could be 50 (Reed and Bryant 2000) because both the low evolutionary potential and the heavy inbreeding depression in some endemic species resulted in extinction (He, Rao, You, *et al.* 2000).

There are several examples of disappearing plant species due to loss of evolutionary potential in history—for example, species of *Anchusa* such as *A. crispa*, *A. sardoa*, and *A. littorea* from the coastal areas of Sardinia (Itlay) (Fiori 1926; Moris 1859; Valsecchi 1976). The genetic data analysis suggests that the two existing species, *A. sardoa* and *A. littorea*, are in the 'Critically Endangered' category because of their reduced evolutionary fitness and they, too, may disappear if further changes in the environment occur (Coppi, Mengoni, and Selvi 2008).

Apart from population genetics, the majority of investigations, however, ignored seed bank genetics, which also plays a significant role in the evolution of important changes in population. The research focusing on seed bank genetics includes the study on reducing genetic losses (Tonsor, Kalisz, and Fischer 1993; Cabin 1996; McCue and Holtsford 1998; Koch, Huthmann, and Bernhardt 2003). In polluted or disturbed sites, the shifts in allelic frequencies may result from selection at loci, which is important for survival in a polluted environment. The loss of genetic variation may be permanent and the population cannot recover, as illustrated in Figure 1 (Deiter 1993; Belfiore and Anderson 1998; Bickham, Sandhu, Hebert, *et al.* 2000).

According to McNeely, Miller, Reid, *et al.* (1990), both inbreeding and loss of genetic diversity have an effect on the extinction of a species, but inbreeding depression has the most instantaneous effect compared to other genetic losses associated with environmental changes or evolutionary potential. Vilas, Miguel, Amaro, *et al.* (2005) also observed that plant species in Spain are affected by inbreeding depression most immediately but there are no instant impact on loss of genetic diversity.

Loss of S-allele at Self-incompatibility Loci

In higher plants, the pollen-pistil interaction mechanism has been found to inhibit self-fertilization, causing a phenomenon called homomorphic self-incompatibility (SI) (DeNettancourt 2001). Self-incompatibility is mechanically controlled by a single S locus (which may contain several genes) (Silva and Goring 2001). The SI specificity

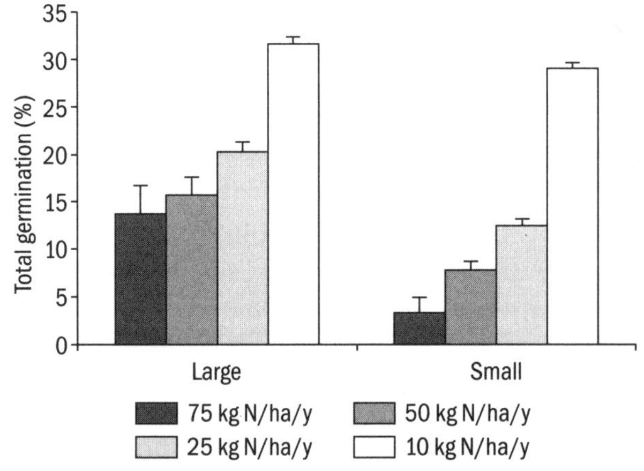

Figure 1 *Effect of eutrophication on germination*
Source *Vergeer, Rengelink, Ouborg, et al. (2003)*

of pollen is dependent on their respective diploid parents (Hiscock and McInnis 2003). Individuals carrying rare, large number of alleles on S loci have more mating opportunities (Schierup 1998).

The low level of genetic diversity partially results in a decline in population, which significantly correlates to population fitness, especially in self-incompatible species (Booy, Groenendael, Hendriks, *et al.* 2000). Similar results were predicted in some endemic plant species in Xinjing (China) (Leimu, Mutikainen, Koricheva, *et al.* 2006). Thus, low levels of genetic diversity may further reduce the population of vulnerable species. Further studies of S-locus (which is responsible for gametophytic self-incompatibility) within populations provide more insights (Tahan, Geng, Zeng, *et al.* 2009). *Hypochaeris salzmanniana* is a species endemic to South America. The analysis of its reproductive biology showed that the loss of S-allele is the main factor for its reduced evolutionary potential, which shows that it is self-incompatible (Ortiz, Tremetsberger, Talavera, *et al.* 2007). In several other studies of Asteraceae, the loss of S-alleles and loss of control of self-incompatibility are deemed responsible for low fruiting or seed set in species such as *Hypochaeris maculate* (Wells 1976), *Aster furcatus* (Reinartz and Les 1994), *Scalesia affinis* (Nielsen, Siegismund, and Philipp 2003), and *Hymenopsis acaulis* var. glabra (DeMauro 1993), and *Incarvillea species* (Qu, Li, Luo, *et al.* 2007; Han, Dai, Yang, *et al.* 2008). The loss in genetic diversity reduces the evolutionary potential of vulnerable plant species, especially through the loss of S-allele in self-incompatible species.

COMPARISON OF RARE AND WIDESPREAD SPECIES

Comparison of widespread species with restricted congeneric species has gained much attention in population genetics, and conservation biology (Karron 1987; Karron, Linhart, Chaulk, *et al.* 1988; Gitzendanner and Soltis 2000; He, Rao, You, *et al.* 2000; Sales, Nebauer, Mus, *et al.* 2001; Burne, Yates, and Ladd 2003; Van Rossum, De Sousa, and Triest 2004; Zietsman, Dreyer, and Vuuren 2009). The comparative studies of genetic variation between rare and widespread species have shown that both these are polymorphic at the same level, as shown in Figure 2 (Gitzendanner and Soltis 2000; Dodd and Helenurm 2002). There are several reasons behind this scenario of species extinction but a few of them are quoted here. For example, small population size, restricted distributions reproductive methodology (Tansely and Brown 2000). Some authors pointed out that the reason behind high genetic variation among rare species was mostly because these are dioeceous and avoid inbreeding depression. Secondly, historical distribution of endangered species is most likely attributed to multiple causes rather than a single event (Chang, Choi, Kim, *et al.* 2007).

Another possible factor is the hybridization of two closely related species or hybridization of two or more related taxa or widespread taxa (Godt and Hamrick 1996; Colosi and Schaal 1997; Wolfe and Elisens 1993; Rieseberg, Fossen, and Van 1995; Smith and Pham 1996). The possible reason behind the high level of genetic variability in endemic species is possibly due to high genetic identity between endemic and widespread species, but only this factor is insufficient (Crawford 1983; Rieseberg and Soltis 1991; Rieseberg 1995). But the high genetic diversity in some endemic species suggests straightforward conservational goals because genetic diversity is positively associated with the ability to be protected from different diseases and stresses (Archibald, Wolf, Tepedino, *et al.* 2001).

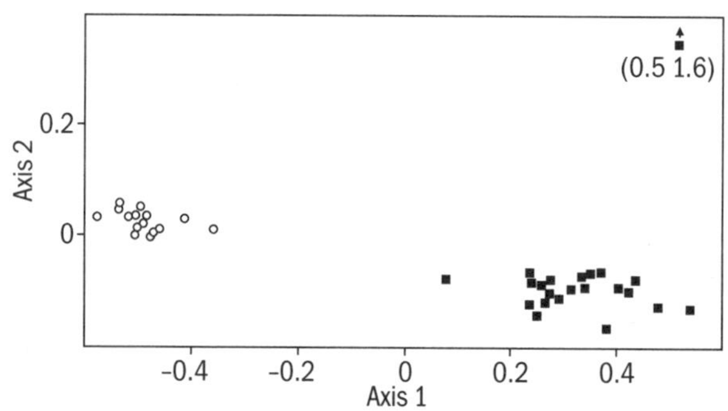

Figure 2 *Genetic variation between widespread and rare plant species (both kinds of species show same level of polymorphism)*

Source *Runyeon and Honor (1996)*

IMPLICATIONS FOR CONSERVATION

Besides genetic erosion, other threats such as collection of particular species for a particular use, overgrazing, uprooting by rodents, and intense collection of wild herbal plants should be minimized in the wild. Especially the medicinal plant supply should be restricted through commercial cultivation only. If the reason behind the reduction of population size and genetic losses is mainly human-induced habitat fragmentation, then the simplest means of genetic preservation is to reduce deforestation and protect the habitats (Zhang, Li, and Qiu 2006). The species used by humans for diverse purposes are most vulnerable and require immediate conservation. Thus, artificial propagation is the recommended method for conservation along with in-situ conservation. This requires a detailed reproductive biology study as a first step to carry out successful conservation management (Cao, Yao, Ding, *et al.* 2006).

The maintenance of genetic diversity is a major objective within conservation plans because genetic diversity is crucial for maintaining evolutionary potential (Hamrick and Godt 1996). The valuable data obtained through basic genetic study mainly contributes in developing conservation plans or in optimizing artificial propagation of endangered species. The information obtained on genetic variation acts as a baseline for an effective conservation management plan. The low level of genetic variation is mostly due to past evolutionary process and the immediate losses are due to human activities. Therefore, the major goal of conservation implications lies in the maintenance of genetic diversity and the process of evolution to avoid extinction (Maccue-Carson 1990; Falk and Holsinger 1991; Godt and Hamrick 1996). The loss of genetic diversity consequently reduces species capacity to cope with environmental and geographical fluctuations in short and long terms (Ellstrand and Elam 1993; Milligan, Leebens-Mack, and Strand 1994; Reisch, Poschlod, and Wingender 2003), and this knowledge plays a significant role in conservation management (Falk and Holsinger 1991; Francisco-Ortega, Santos-Guerra, Kim, *et al.* 2000).

In genetic consideration, gene flow is the indicator of micro-evolutionary force (Slatkin 1987). Therefore, limited gene flow in isolated and fragmented population may result in high inbreeding depression. Thus, the loss of rare alleles could possibly be expected. For the long-term survival of such species by in-situ conservation, the immediate need and extended programe is to restore a large area to minimize loss in genetic diversity and genetic drift in future. With regard to existing conservation measures, artificial cultivation of as many species as possible should be designed, which may include representatives of the whole population (Ellstrand and Elam 1993; Frankham, Ballou, and Briscoe 2002; Chen, Wang, Chen, *et al.* 2009). When it is difficult to conserve a plant by ex-situ conservation, the in-situ conservation is the best alternative (Bawa and Ashton 1991; He, Rao, You, *et al.* 2000).

Mixed population translocation programmes with monitoring for outbreeding depression are thought to be the most successful in the long term. For example, a nursery of a stock of plants from all possible populations and also translocation of small and young plants, vulnerable in a particular habitat may reduce the effects of inbreeding (Kingston and Waldren 2003).

There is an immense scope to use endangered but architectural species around settlements for the assurance of their survival. Apart from this, propagation courses are provided to the natives of that area to enhance their expertise to maintain nursery and supporting recovery of species (Kingston, Waldren, and Smyth 2004). Similarly, replacement by transplanting seedlings from established populations of lost populations at specific points could also be possible where habitats are severely affected by different factors causing loss in genetic diversity (Wang, Li, Hao, *et al.* 2004; Wang, Zhang, Qian, *et al.* 2008).

Some authors argued that introduced population established from mixed propagules and sources could not serve as a source for the restoration of threatened species. Though mixed genotype may prove a better strategy, it may reduce the ability of the reintroduced population to maintain itself and may lead to outbreeding depression. To minimize these kinds of effects, the seeds collected from a specific natural population should be planted in a special area without gene flow as a first step, and then reintroduced into natural population (Barrett and Kohn 1991; Waser 1993; Li, Chen, Zhang, *et al.* 2005;

Li and Chen 2009). If a population suffers from loss of fitness and high inbreeding depression, then their impact could be minimized by restoring a viable seed bank (Buza, Young, and Thrall 2000).

Generally, the botanic gardens are playing an important role in preserving rare and endangered species (Maunder 1994a, b; Li, Xu, and He 2002). However, there are some disadvantages of botanic gardens such as space limitations and the growing of a few individuals and ignoring a wider range of species (Li, Xu, and He 2002).

CONCLUSION

Among genetic factors, inbreeding depression has more pronounced consequences on fitness for the survival of the species. The loss of genetic diversity has the impact of reducing the evolutionary potential of vulnerable plant species, especially the loss of Sallele in self-incompatible species has a lethal effect. Further, restricted distribution of species could be due to more genetic variation and low population size, which could lead to their increased vulnerability. The implications of conservation for these vulnerable species should cover detailed research of all contributing factors and then ultimately develop a multi-perspective strategy for their restoration/ conservation.

REFERENCES

Archibald, K. J., G. P. Wolf, J. V. Tepedino, and J. Bair. 2001. Genetic relationships and population structure of the endangered steamboat buckwheat, *Eriogonum ovalifolium* var. *williamsiae* (Polygonaceae). *American Journal of Botany* 88(4):608–615

Astorga, G. J., P. A. Vovides, D. Toledo, and F. Morejon. 2009. Diversity and genetic structure of the endangered cycad *Dioon sonorense* (Zamiaceae) from Sonora, Mexico: evolutionary and conservation implications. *Biochemical Systematics and Ecology* 36:891–899

Barrett, S. C. H., and J. R. Kohn. 1991. Genetic and evolutionary consequences of small population size in plants: implications for conservation. In *Genetics and Conservation of Rare Plants*, edited by D. A. Falk and K. E. Holsinger, pp. 3–30. New York: Oxford University Press

Bawa, K. S., and P. S. Ashton. 1991. Conservation of rare trees in tropical rain forest: a genetic perspective. In *Genetics and Conservation of Rare Plants*, edited by D. A. Falk and K. E. Holsinger, pp. 62–71. Oxford: Oxford University Press

Belfiore, N. M., and S. L. Anderson. 1998. Genetic patterns as a tool for monitoring and assessment of environmental impacts: the example of genetic ecotoxicology. *Environmental Monitoring Assessment* 51:465–479

Bickham, W. J., S. Sandhu, N. D. P. Hebert, L. Chikhi, and R. Athwal. 2000. Effects of chemical contaminants on genetic diversity in natural populations: implications for biomonitoring and ecotoxicology. *Mutation Research* 463:33–51

Booy, G., J. Groenendael, R. Hendriks, M. Smulders, and B. Vosman. 2000. Genetic diversity and the survival of populations. *Plant Biology* 2:379–395

Brook, B. W., D. W. Tonkyn, J. J. O'Grady, and R. Frankham. 2002. Contribution of inbreeding to extinction risk in threatened species. *Conservation Ecology* 6(1):16. Details available at <www.consecol.org/vol16/iss11/art16>

Burne, H. M., C. J. Yates, and P. G. Ladd. 2003. Comparative population structure and reproductive biology of the critically endangered shrub *Grevillea althoferorum* and two closely related more common congeners. *Biological Conservation* 114:53–65

Buza, L., A. Young, and P. Thrall. 2000. Genetic erosion, inbreeding and reduced fitness in fragmented populations of the endangered tetraploid pea *Swainsona recta*. *Biological Conservation* 93:177–186

Cabin, R. J. 1996. Genetic comparison of seed bank and seedling populations of the desert mustard *Lesquerella fendleri*. *Evolution* 50:1830–1841

Cao, P. J., Q. F. Yao, B. Y. Ding, H. Y. Zeng, Y. X. Zhong, C. X. Fu, and X. F. Jin. 2006. Genetic diversity of *Sinojackia dolichocarpa* (Styracaceae), a species endangered and endemic to China, detected by inter-simple sequence repeat (ISSR). *Biochemical Systematics and Ecology* 34:231–239

Chang, C. S., D. Y. Choi, H. Kim, Y. S. Kim, and T. Y. Park. 2007. Mating system and genetic diversity in the threatened plant, *Kirengeshoma palmata* (Saxifragaceae) in Korea. *Journal of Plant Research* 120:149–156

Chang, S. C., H. Kim, Y. T. Park, and M. Maunder. 2004. Low levels of genetic variation among southern peripheral populations of the threatened herb, *Leontice microrhyncha* (Berberidaceae) in Korea. *Biological Conservation* 119:387–396

Chen, F., A. Wang, K. Chen, D. Wan, and J. Liu. 2009. Genetic diversity and population structure of the endangered and medically important *Rheum tanguticum* (Polygonaceae) revealed by SSR Markers. *Biochemical Systematics and Ecology* 37(5):613–621

Chen, J. M., J. Y. Wang, X. Liu, Y. W. Zhang, and Q. F. Wang. 2004. RAPD analysis for genetic diversity of *Isoetes sinensis*. *Biodiversity Science* 12:348–353

Chen, M. J., X. Liu, Y. J. Wang, W. G. Robert, and F. Q. Wang. 2005. Genetic variation within the endangered quillwort *Isoetes hypsophila* (Isoetaceae) in China as evidenced by ISSR analysis. *Aquatic Botany* 82:89–98

Chen, Y. Y., L. X. Li, Y. L. Yin, and W. Li. 2008. Genetic diversity of the threatened aquatic plant *Ottelia alismoides* in the Yangtze River. *Aquatic Botany* 88:10–16

Colosi, J. C., and B. A. Schaal. 1997. Wild proso millet (*Panicum miliaceum*) is genetically variable and distinct from crop varieties of proso millet. *Weed Science* 45: 509–518

Coppi, A., A. Mengoni, and F. Selvi. 2008. AFLP fingerprinting of *Anchusa* (Boraginaceae) in the Corso-Sardinian system: genetic diversity, population differentiation and conservation priorities in an insular endemic group threatened with extinction. *Biological Conservation* 141:2000–2011

Craig, J. L. 1994. Meta-populations: is management as flexible as nature? In *Creative Conservation: Interactive Management of Wild and Captive Animals*, edited by P. J. S. Olney, G. M. Mace, and A. T. C. Feistner, pp. 50–66. London: Chapman and Hall

Crawford, D. J. 1983. Phylogenetic and systematic inferences from electrophoretic studies. In *Isozymes in Plant Genetics and Breeding, Part A*, edited by S. D. Tanksley and T. J. Orton, pp. 257–287. Amsterdam: Elsevier Science Publishers

Deiter, M. P. 1993. Identification and quantification of pollutants that have the potential to affect evolutionary processes. *Environmental Health Perspectives* 101:278

DeMauro, M. M. 1993. Relationship of breeding system to rarity in the Lakeside daisy (*Hymenoxys acaulis* var. Glabra). *Conservation Biology* 7: 542–550

DeNettancourt, D. 2001. *Incompatibility and Incongruity in Wild and Cultivated Plants*. Berlin: Springer-Verlag

Dodd, S. C., and K. Helenurm. 2002. Genetic diversity in *Delphinium variegatum* (Ranunculaceae): a comparison of two insular endemic subspecies and their widespread mainland relative. *American Journal of Botany* 89:613–622

Eckstein, R. L., R. A. Oneill, J. Danihelka, A. Otte, and W. Kohler. 2006. Genetic structure among and within peripheral and central populations of three endangered floodplain violets. *Molecular Ecology* 15:2367–2379

Elgar, M. A., and D. Clode. 2001. Inbreeding and extinction in island populations: a cautionary tale. *Conservation Biology* 15:284–286

Ellstrand, N. C., and D. R. Elam. 1993. Population genetic consequences of small population size: implications for plant conservation. *Annual Review of Ecology and Systematics* 24:217–242

Falk, D. A., and K. E. Holsinger (eds). 1991. *Genetics and Conservation of Rare Plants*. New York: Oxford University Press

Fenster, C. B., and L. F. Galloway. 2000. Inbreeding and outbreeding depression in natural populations of *Chamaecrista fasciculata* (Fabaceae). *Conservation Biology* 14:1406–1412

Fiori, A. 1926. Nuova flora analitica d'Italia Ricci. *Firenze* 2:280–283

Fontdevila, A., and A. Moya. 1999. *Introducción a la Genética de Poblaciones*. Madrid, Editorial Síntesis

Francisco-Ortega, J., A. Santos-Guerra, S. C. Kim, and D. J. Crawford. 2000. Plant genetic diversity in the Canary Islands: a conservation perspective. *American Journal Botany* 87:909–919

Frankel, O. H., and M. E. Soule. 1981. *Conservation and Evolution*. New York: Cambridge University Press

Frankham, R., and J. G. Kingsolver. 2004. Response to environmental change: adaptation or extinction. In *Evolutionary Conservation Biology*, edited by R. Ferrier, U. Dieckman, and D. Couvet, pp. 85–100. New York: Cambridge University Press

Frankham, R. 1996. Relationship of genetic variation to population size in wildlife. *Conservation Biology* 10:1500–1508

Frankham, R. 2005. Genetics and extinction: a review. *Biological Conservation* 126:131–140

Frankham, R., J. D. Ballou, and D. A. Briscoe. 2002. *Introduction to Conservation Genetics*. UK: Cambridge University Press

Gapare, J. W., S. N. Aitken, and C. E. Ritland. 2005. Genetic diversity of core and peripheral Sitka spruce (*Picea sitchensis* [Bong.] Carr) populations: implications for conservation of widespread species. *Biological Conservation* 123:113–123

Gaston, K. J., and W. E. Kunin. 1997. Concluding comments. In *The Biology of Rarity*, edited by W. E. Kunin and K. J. Gaston, pp. 262–272. London: Chapman and Hall

Gitzendanner, M. A., and P. S. Soltis. 2000. Patterns of genetic variation in rare and widespread congeners. *American Journal of Botany* 87:1513–1520

Glendinning, D. 1989. Some aspects of autotetraploid population dynamics. *Theoretical and Applied Genetics* 78:233–242

Godt, M. J. W., and J. L. Hamrick. 1996. Genetic diversity and morphological differentiation in *Liatris helleri* (Asteraceae), a threatened plant species. *Biodiversity and Conservation* 5:461–471

Gray, A. 1996. Genetic diversity and its conservation in natural populations of plants. *Biodiversity Letters* 3:71–80

Groombridge, J. J., C. G. Jones, M. W. Bruford, and R. A. Nichols. 2000. Ghost alleles of the Mauritius kestrel. *Nature* 403:616

Hamrick, J. L., and M. J. Godt. 1996. Conservation genetics of endemic plant species. In *Conservation Genetics*, edited by J. C. Avise and J. L. Hamrick, pp. 281–304. New York: Chapman and Hall

Han, Y., C. Dai, C. F. Yang, Q. F. Wang, and T. J. Motley. 2008. Anther appendages of *Incarvillea* trigger a pollen-dispensing mechanism. *Annals of Botany* 102:473–479

Hartl, D. L., and A. G. Clark. 1989. *Principles of Population Genetics*, 2nd edn. Massachusetts: Sinauer Associates

Hartl, D. L., and G. C. Clark. 1997. *Principles of Population Genetics*, 3rd edn. Massachusetts: Sinauer Associates

He, T., G. Rao, R. You, S. Ge, and D. Zhang. 2000. Genetic diversity of widespread *Ophiopogon intermedius* (Liliaceae s.l.): A comparison with endangered *O. xylorrhizus*. *Biological Conservation* 96:253–257

Hedrick, P. W. 2000. *Genetics of Populations*, 2nd edn. Sudbury: Jones and Bartlett Publishers

Hiscock, S. J., and S. M. McInnis. 2003. Pollen recognition and rejection during the sporophytic self-incompatibility response: Brassica and beyond. *Trends Plant Science* 8: 606–613

Hsu, T. W., S. J. Moore, and T. Y. Chiang. 2000. Low RAPD polymorphism in *Archangiopteris itoi*, a rare and endemic fern in Taiwan. *Botanical Bulletin of Academia Sinica* 41:15–18

Huenneke, L. F. 1991. Ecological implications of genetic variation in plant populations. In *Genetics and Conservation of Rare Plants*, edited by D. A. Falk and K. E. Holsinger. New York: Oxford University Press

Husband, B., and D. Schemske. 1997. The effect on inbreeding in diploid and tetraploid populations of *Epilobium angustifolium* (Ona-graceae): implications for the genetic basis of inbreeding depression. *Evolution* 51:737–746

Kadereit, G., and W. J. Kadereit. 2005. Phylogenetic relationships, evolutionary origin, taxonomic status, and genetic structure of the endangered local Lower Elbe River (Germany) endemic *Oenanthe conioides* (Nolte ex Rchb.f.) Lange (Apiaceae): ITS and AFLP evidence. *Flora* 200:15–29

Karron, J. D. 1987. A comparison of levels of genetic polymorphism and self-compatibility in geographically restricted and widespread plant congeners. *Evolutionary Ecology* 1:47–58

Karron, J. D., Y. B. Linhart, C. A. Chaulk, and C. A. Robertson. 1988. Genetic structure of populations of geographically restricted and widespread species of *Astragalus* (Fabaceae). *American Journal of Botany* 75(8):1114–1119

Kingston, N., and S. Waldren. 2003. The plant communities and environmental gradients of Pitcairn Island: the significance of invasive species and the need for conservation management. *Annals of Botany* 92: 31–40

Kingston, N., S. Waldren, and N. Smyth. 2004. Conservation genetics and ecology of *Angiopteris chauliodonta* Copel. (Marattiaceae), a critically endangered fern from Pitcairn Island, South Central Pacific Ocean. *Biological Conservation* 117:309–319

Knapen, D., G. Knaepkens, L. Bervoets, M. I. Taylor, M. Eens, and E. Verheyen. 2003. Conservation units based on mitochondrial and nuclear DNA variation among European bullhead populations (*Cottus gobio*) from Flanders (Belgium). *Conservation Genetics* 4:129–140

Koch, M., M. Huthmann, and K. G. Bernhardt. 2003. *Cardamine amara* L. (Brassicaceae) in dynamic habitats: genetic composition and diversity of seed bank and established populations. *Basic and Applied Ecology* 4:339–348

Lande, R. 1995. Mutation and conservation. *Conservation Biology* 9:783–791

Landergott, U., R. Holderegger, G. Kozlowski, and J. J. Schneller. 2001. Historical bottlenecks decrease genetic diversity in natural populations of *Dryopteris cristata*. *Heredity* 87:344–355

Leimu, R., P. Mutikainen, J. Koricheva, and M. Fischer. 2006. How general are positive relationships between plant population size, fitness and genetic variation? *Journal of Ecology* 94:942–952

Li, H., and G. Chen. 2009. Genetic variation within the endangered mangrove species *Sonneratia paracaseolaris* (Sonneratiaceae) in China detected by inter-simple sequence repeats analysis. *Biochemical Systematics and Ecology* 37:260–265

Li, Q., Z. Xu, and T. He. 2002. Ex situ genetic conservation of endangered *Vatica guangxiensis* (Dipterocarpaceae) in China. *Biological Conservation* 106: 151–156

Li, Y. Y., X. Y. Chen, X. Zhang, T. Y. Wu, H. P. Lu, and Y. W. Cai. 2005. Genetic differences between natural and introduced populations of *Metasequoia glyptostroboides*: implications for species recovery. *Conservation Biology* 19: 224–231

Liu, P., S. Y. Yang, Y. C. Hao, and D. W. Guo. 2007. Ecological risk assessment using RAPD and distribution pattern of a rare and endangered species. *Chemosphere* 68:1497–1505

Lynch, M. 1991. The genetic interpretation of inbreeding depression and outbreeding depression. *Evolution* 45:622–629

Lynch, M. 1996. A quantitative genetic perspective on conservation issues. In *Conservation Genetics: Case Histories from Nature*, edited by J. C. Avise and J. L. Hamrick, pp. 471–501. New York: Chapman and Hall

Lynch, M., and R. Lande. 1998. The critical effective size for a genetically secure population. *Animal Conservation* 1:70–72

Maccue-Carson, H. L. 1990. Increasing genetic variance after a population bottleneck. *Trends in Ecology & Evolution* 5:228–231

Maunder, M. 1994a. Botanical gardens: future challenges and responsibilities. *Biodiversity and Conservation* 3:97–103

Maunder, M. 1994b. Practical aspects of plant conservation in a botanic garden, the relationship between botanic gardens and the wild habitat. *Biossiera* 47:155–165

McCue, K. A., and T. P. Holtsford. 1998. Seed bank influences on genetic diversity in the rare annual *Clarkia springvillensis* (Onagraceae). *American Journal of Botany* 85:30–36

McNeely, J. A., K. R. Miller, W. V. Reid, R. A. Mittermeier, and T. B. Werner. 1990. *Conserving the World's Biological Diversity*. Washington, DC: WWF-US and the World Bank

Milligan, B. G., J. Leebens-Mack, and A. E. Strand. 1994. Conservation genetics: beyond the maintenance of marker diversity. *Molecular Ecology* 3:423–435

Moris, G. G. 1859. Flora Sardoa: Ex regio typographeo. *Torino* 3:139–143

Newman, D., and D. Pilson. 1997. Increased probability of extinction due to decreased genetic effective population size: experimental populations of *Clarkia pulchella*. *Evolution* 51:354–362

Nielsen, L. R., H. R. Siegismund, and M. Philipp. 2003. Partial self-incompatibility in the polyploid endemic species *Scalesia affinis* (Asteraceae) from the Galápagos: remnants of a self-incompatibility system? *Botanical Journal of Linnean Society* 142:93–101

Ortiz, M. Á., K. Tremetsberger, S. Talavera, T. F. Stuessy, and J. L. García-Castaño. 2007. Population structure of *Hypochaeris salzmanniana* DC. (Asteraceae), an endemic species to the Atlantic coast on both sides of the Strait of Gibraltar, in relation to Quaternary sea level changes. *Molecular Ecology* 16:541–552

Primack, R. B. 1993. *Vulnerability to Extinction: Essentials of Conservation Biology*. Massachusetts: Sinauer Associates

Qu, R. M., X. J. Li, Y. B. Luo, M. Dong, H. L. Xu, X. Chen, and A. Dafni. 2007. Wind-dragged corolla enhances self-pollination: a new mechanism of delayed self-pollination. *Annals of Botany* 100:1155–1164

Quilichini, A., D. J. Thompson, and M. Debussche. 2001. Evidence for local outbreeding depression in the Mediterranean island endemic *Anchusa crispa* Viv. Boraginaceae. *Heredity* 87:190–197

Quilichini, A., M. Debussche, and D. J. Thompson. 2004. Geographic differentiation of morphological traits and isozymes in the Mediterranean island endemic *Anchusa crispa*: implications for the conservation of a protected species. *Biological Conservation* 118:651–660

Reed, D. H., and E. H. Bryant. 2000. Experimental tests of minimum viable population size. *Animal Conservation* 3:7–14

Reinartz, J. A., and D. H. Les. 1994. Bottleneck-induced dissolution of self-incompatibility and breeding system consequences in *Aster furcatus* (Asteraceaae). *American Journal Botany* 81:446–55

Reisch, C., P. Poschlod, and R. Wingender. 2003. Genetic variation of *Saxifraga paniculata* Mill. (Saxifragaceae): molecular evidence for glacial relict endemism in Central Europe. *Biological Journal of Linnean Society* 80:11–21

Rieseberg, L. H. 1995. The role of hybridization in evolution: old wine in new skins. *American Journal of Botany* 82:944–953

Rieseberg, L. H., and D. E. Soltis. 1991. Phylogenetic consequences of cytoplasmic gene flow in plants. *Evolutionary Trends in Plants* 5:65–84

Rieseberg, L. H., C. Fossen, and C. Van. 1995. Hybrid speciation accompanied by genomic reorganization in wild sunflowers. *Nature* 375:313–316

Runyeon, H., and C. Honor. 1996. Genetic structure in the species-pair *Silene vulgaris* and *S. uniflora* (Caryophyllaceae) on the Baltic Island of Öland. *Ecography* 19(2):181–193

Saccheri, I., M. Kuussaari, M. Kankare, P. Vikman, W. Fortelius, and I. Hanski. 1998. Inbreeding and extinction in a butterfly metapopulation. *Nature* 392:491–494

Sales, E., S. G. Nebauer, M. Mus, and J. Segura. 2001. Population genetic study in the Balearic endemic plant species *Digitalis minor* (Scrophulariaceae) using RAPD markers. *American Journal of Botany* 88(10):1750–1759

Schierup, M. H. 1998. The number of self-incompatibility alleles in a finite, subdivided population. *Genetics* 149:1153–1162

Sherwin, W. B., and C. Moritz. 2000. Managing and monitoring genetic erosion. In *Genetics, Demography and Viability of Fragmented Populations*, edited by G. M. Clarke and A. G. Young, pp. 9–34. UK: Cambridge University Press

Silva, N. F., and D. R. Goring. 2001. Mechanisms of self-incompatibility in flowering plants. *Cell & Molecular Life Sciences* 58:1988–2007

Slatkin, M. 1987. Gene flow and the geographic structure of populations. *Science* 236:787–792

Smith, J. F., and T. V. Pham. 1996. Genetic diversity of the narrow endemic *Allium aaseae* (Alliaceae). *American Journal of Botany* 83:717–726

Soule, M. E., and G. H. Orians (ed.). 2001. *Conservation Biology: Research Priorities for the Next Decade*. Washington, DC: Island Press

Tahan, O., Y. Geng, L. Zeng, S. Dong, F. Chen, J. Chen, Z. Song, and Y. Zhong. 2009. Assessment of genetic diversity and population structure of Chinese wild almond, *Amygdalus nana*, using EST and genomic SSRs. *Biochemical Systematics & Ecology* 37: 146–153

Tansley, A., and C. R. Brown. 2000. RAPD variation in the rare and endangered *Leucadendron elimense* (Proteaceae): implications for their conservation. *Biological Conservation* 95:39–48

Tonsor, S. J., S. Kalisz, and J. Fischer. 1993. A life history based study of population structure: seed bank to adults in *Plantago lanceolata*. *Evolution* 47:833–843

Valsecchi, F. 1976. Il genere *Anchusa* in Sardegna. *Webbia* 30(1):43–68

Van Rossum, F., S. C. De Sousa, and L. Triest. 2004. Genetic consequences of habitat fragmentation in an agricultural landscape on the common *Primula veris*, and comparisons with its common rare congener, *P. vulgaris*. *Conservation Genetics* 5:231–245

Vergeer, P., R. Rengelink, J. N. Ouborg, and M. G. J. Roelofs. 2003. Effects of population size and genetic variation on the response of *Succisa pratensis* to eutrophication and acidification. *Journal of Ecology* 91(4):600–609

Vilas, C., E. S. Miguel, R. Amaro, and C. Garcia. 2005. Relative contribution of inbreeding depression and eroded adaptive diversity to extinction risk in small populations of shore campion. *Conservation Biology* 20(1):229–238

Wang, C., H. Zhang, Z. Q. Qian, and G. F. Zhao. 2008. Genetic differentiation in endangered *Gynostemma pentaphyllum* (Thunb.) Makino based on ISSR polymorphism and its implications for conservation. *Biochemical Systematics and Ecology* 36:699–705

Wang, D. L., Z. C. Li, G. Hao, T. Y. Chiang, and X. J. Ge. 2004. Genetic diversity of *Calocedrus macrolepis* (Cupressaceae) in south-western. *Biochemical Systematics and Ecology* 32:797–807

Waser, N. M. 1993. Population structure, optimal outbreeding, and assortative mating in angiosperms. In *The Natural History of Inbreeding and Outbreeding, Theoretical and Empirical Perspectives*, edited by N. W. Thornhill, pp. 173–199. Chicago and London: University of Chicago Press

Wells, T. C. E. 1976. *Hypochaeris maculata* L. (*Achyrophorus maculatus* [L.] Scop.). *Journal of Ecology* 64:757–774

Wolfe, A. D., and W. J. Elisens. 1993. Diploid hybrid speciation in Penstemon (Scrophulariaceae) revisited. *American Journal of Botany* 80:1082–1094

Ye, Q. G., S. Y. Wang, H. Z. Xu, and C. Wang. 1996. A preliminary study on the conservative status of *Sinojackia dolichocarpa*. Chin. *Biodiversity* 4:135–138

Young, A., T. Boyle, and T. Brown. 1996. The population genetic consequences of habitat fragmentation for plants. *Trends in Ecology & Evolution* 11:413–418

Zhang, X. P., X. H. Li, and Y. X. Qiu. 2006. Genetic diversity of the endangered species *Kirengeshoma palmata* (Saxifragaceae) in China. *Biochemical Systematics and Ecology* 34:38–47

Zietsman, J., L. L. Dreyer, and J. V. Vuuren. 2009. Genetic differentiation in Oxalis (Oxalidaceae): a tale of rarity and abundance in the Cape Floristic Region. *South African Journal of Botany* 75:27–33

CHAPTER 8

Use of Geospatial Techniques in Monitoring and Management of Biodiversity of Urban Ecosystems

Noresah Mohd Shariff and Sanjay Gairola

INTRODUCTION

Urban ecosystems are complex and dynamic systems consisting of natural, economic, and social factors. Today urbanization is a reality in the world and it is predicted that the trend of rapid urbanization will bring with it a number of challenges at the local, regional, and global levels (Chen, Adimo, and Bao 2009). The biological diversity of urban areas is heavily influenced by the built environment and the economic, social, and cultural dynamics. Urbanization, land and landscapes, climate change, and biodiversity loss are the top priorities in the current scientific research worldwide. The negative impact of urban growth on environment has been demonstrated in the developed and developing countries (Chiesura 2004; Colding 2007; Theobald, Hobbs, Bearly, *et al.* 2000; Tzoulas, Korpela, Venn, *et al.* 2007). The loss and degradation of global biodiversity due to rapid urbanization is manifest because land transformation disrupts key ecosystem functions, making them more susceptible to change and weakening their capability to deliver key ecosystem goods and services to humans. Urbanization has profoundly transformed natural landscapes throughout the world and caused irreversible effects on the structure, function, and dynamics of ecological systems across a wide range of spatial scales (Luck and Wu 2002; Beardsley, Thorne, Roth, *et al.* 2009).

Research on the patterns and mechanisms of biodiversity could increase our ability to understand, manage, and mitigate the negative effects of human expansion on biological resources (Hansen, Knight, Marzluff, *et al.* 2005). Urban systems are considered complex ecosystems, which require integrated spatial and temporal modelling approaches to better understand the underlying patterns and processes. Along with traditional biodiversity assessment and management approaches, geospatial tools currently provide robust ways to map and monitor urban biodiversity. Researchers emphasize that geospatial tools can provide timely and extensive spatial data from which urban ecosystem attributes such as land cover, forest structure, species composition and condition, heat island effects, and carbon storage can be derived (Ward and Johnson 2007). The urban plant communities are already strongly affected by anthropogenic influences such as disturbance and the urban heat island effect (Honour, Bell, Ashenden, *et al.* 2009).

There is mounting evidence that the geographical expansion of urban areas and land-use and land-cover changes have a negative effect on the original biological diversity in the urban areas (Grimm, Faeth, Golubiewski, *et al.* 2008). Capturing the trends and status of urban biodiversity is challenged by the current rapid changes, including increasing flow of population into urban areas and multiple pressures on neighbouring ecosystems (Kohsaka 2010). Until recently, only a small percentage of ecologists studied cities (Bryant 2006); researchers focused on the ecology of more pristine wilderness areas where the effects of humans were minimal. However, this situation is changing as the ecological effects of unprecedented amounts of land consumption in urban areas and loss of native species due to urbanization are recognized. Many recent studies have highlighted the use of spatial metrics and geostatistics in urban analysis and geographically weighted regression to connect urban form and urban growth factors. Others use geospatial tools such as remote sensing (RS) and Geographic Information System (GIS) to map and identify spatial variability in urban structures, exploring the correlation between population density and urban form and spatial modelling for sustainable urban planning.

The 2010 biodiversity target of halting biodiversity loss is a milestone to promote applied research on urban ecology, biodiversity, and the sustainable use of ecosystem goods and services. Consequently, most of the research broadly emphasizes sustainable urban development and maximizes the ability of urban ecosystems to support biodiversity. Conservationists increasingly recognize the potential of managing human-inhabited landscapes for biodiversity protection and enhancement because healthy and diverse ecosystems are the foundation for human well-being as well. Sustainable development requires balancing economic development with the need to improve and sustain the quality of life, conserve resources, and coexist in harmony with nature, thereby protecting future generations as well. As geography is a basic element in this quest, geospatial technology plays a key role in ensuring informed decisions aimed at creating a sustainable environment. In general, geospatial technology is finding increased relevance in almost every walk of life in domains like infrastructure, utilities, natural resource management, disaster and risk management, environmental impact assessment, and many others. The RS and GIS tools have particular relevance in many developing countries due to their potential to improve biodiversity monitoring and assessment processes and guide sustainable urban development. In this chapter, there is emphasis on the utility of RS and GIS tools in urban ecosystem studies, outlining some of the published literature specifically targeted at urban biodiversity assessment and management, and highlighting the potential of geospatial tools for dealing with urban biodiversity issues, including sustainable urban development.

MONITORING URBAN BIODIVERSITY

Biodiversity research on a broader landscape scale has revealed that urban areas can contain a relatively high level of biodiversity. In particular, it is considered that urban forests can contain a significant percentage of species that naturally occur in an area (Alvey 2006).

Most research on urban biodiversity has focused on the diversity within a city or country. However, the processes and patterns that shape biodiversity within and around cities also need to be fully understood (Kinzig, Warren, Martin, *et al.* 2005). The changes in land-use patterns due to urbanization provide many social and economic benefits, but they come at a cost to the natural environment (Tang, Xue, Yuc, *et al.* 2005). Urbanization has an enormous impact on the environment at the local, regional, and global scales, and has many implications for ecosystem functions, which include changes in productivity and soil nutrient availability, modification to structural characteristics of landscapes, extinction of species, and alteration of food and habitats. Monitoring and management of land-use changes with respect to their impact on the various landscape functions (aesthetic, ecological, socio-cultural, and economic) therefore need to be understood and documented adequately (Hobbs 1997; Olaf 2000; Terkenli 2001). Some researchers such as Grimmond and Souch (1994) and Booth and Jackson (1997) highlighted that the implications of urbanization on the environment are significant and include loss of vegetation and open spaces, a decline in the extent and connectivity of wetlands and wildlife habitat, increased temperatures and green house gas emissions, and a reduction in the quality of soil and water. Therefore, integrated research on various human and ecological processes is required for holistic understanding of urban environment (Ellis and Ramankutty 2008).

In the context of urban biodiversity, Alvey (2006) suggests that by studying historical patterns of change, one may be able to predict species that are at the risk of local extinction in a better way, and equipped to establish long-term biodiversity management plans. Similarly, to avoid a reduction in urban biodiversity on a global level (Wittig 1998), using seed material or plant species indigenous to the region for artificially created garden greenery in the city is recommended as far as possible. Considering the high rate of urbanization in developing countries, urban research that focuses on the long-term monitoring of biodiversity, ecosystem services, and human action on shaping urban landscapes has great significance and many environmental implications. In a comprehensive paper, Alberti, Marzluff, Shulenberger, *et al.* (2003) identified four fundamental questions that need to be addressed by urban ecology scholars. These questions are as follows.

1. How do socioeconomic and biophysical variables influence the spatial and temporal distributions of human activities in human-dominated ecosystems?
2. How do spatial and temporal distributions of human activities redistribute energy and material fluxes, and modify disturbance regimes?
3. How do human populations and activities interact with processes at the levels of the population, community, and the individual to determine the resilience of human-dominated systems?
4. How do humans respond to changes in ecological conditions, and how do these responses vary regionally and culturally?

With these considerations, urban ecosystems are presently gaining prime recognition worldwide; subsequently, interest in biodiversity monitoring and sustainable development has increased among researchers. However, some researchers argued that studies on urban biodiversity, especially in multiple habitat types, are a minor focus of conservation biology, though the need for biodiversity conservation in urban ecosystems is being increasingly recognized (McIntyre 2000; McKinney 2002; Miller and Hobbs 2002; Turner, Nakamura, and Dinetti 2004). Some recent studies have examined relationships between biodiversity and urbanization (Turner, Nakamura, and Dinetti 2004; Shochat, Warren, Faeth, *et al.* 2006; Mc Kinney 2008; McIntyre 2000; Yamaguchi 2004; Rango 2005; Sadler, Small, Fiszpan, *et al.* 2006; Smith, Warren, Thompson, *et al.* 2006; Thompson and McLachlan 2007; McIntyre, Rango, Fagan, *et al.* 2001). Among others, the world's largest initiative 'Convention on Biological Diversity' is an international treaty for the conservation of biodiversity, sustainable use of components of biodiversity, and the equitable sharing of the benefits derived from the use of genetic resources. The convention seeks to address all the threats to biodiversity and ecosystem services, including threats from climate change through scientific assessments, development of tools, incentives, and processes; transfer of technologies, and good practices. A landmark initiative, notably the first ever 'City Biodiversity Summit' held in October 2010, at Nagoya, Japan is another substantiation of the increasing awareness of biodiversity conservation in urban areas. This summit aims to confirm the importance of cities and biodiversity, facilitate the sharing of information and exchange of experiences among local authorities worldwide, and appeal to the world to further expand local initiatives to conserve biodiversity.

The modern technology of remote sensing, which includes both aerial and satellite-based systems, allows us to collect a large amount of physical data with ease, speed, and on a repetitive basis. Together with GIS, it helps us analyse the data spatially, offering possibilities of generating various options modelling, and optimizing the whole planning process (Verma, Kumari, and Tiwary 2009). These information systems also offer interpretation of physical (spatial) data together with other socio-economic data, thus providing an important link in the planning process and making it more effective and meaningful. Remote sensing technologies offer advantages over field-based surveys by enabling large spatial coverage and frequent data collection over many inaccessible areas at a relatively low cost. Moreover, remote sensing systems offer a unique and highly flexible tool to survey and monitor biophysical resources and their characterization, and track changes in the composition, extent, and distribution of communities and ecosystems (Wilkie and Finn 1996).

In recent decades, remote sensing has been applied to various urban biodiversity and land-cover analyses, and these applications continue to evolve alongside technological advancements. Moreover, remote sensing imagery can provide information that is well suited to extensive mapping of urban vegetated surfaces. Recent developments in high spatial resolution sensors (for example, <5 m) such as IKONOS and Quickbird have further enabled detailed analysis of urban areas (Johnson, Deshmukh, and Kale 2008). High spatial resolution sensors and laser scanning devices are useful tools for examining urban environments since they can capture detailed information regarding the material and structural composition of the urban surface. The efficacy of the geospatial approach to land-use change analysis and biodiversity management is being increasingly recognized since it can capture detailed and repeated information about the relevant areas.

Using modern tools, a range of recent studies have focused on monitoring urban growth pattern, land-use and land-cover changes, urban green space patterns, and biodiversity conservation in urban areas (Herold, Goldstein, and Clarke 2003; Jim and Chen 2003; Bauer, Yuan, and Sawaya 2003; Yuan, Sawaya, Leoffelholz, *et al.* 2005; Jim and Chen 2006a, 2006b; Jensen and Im 2007; Kazmierczak and James 2008; Faryadi and Taheri 2009; Rafiee, Mahiny, and Khorasani 2009; Miller, Snyder, Skibbe, *et al.* 2009; Chen, Adimo, and Bao 2009). An overview of some recent urban biodiversity studies is given in Table 1.

Geographic Information System tools also provide an additional dimension to data analysis by visualizing the complex patterns and relationships that characterize real-world planning and policy problems in urban environments. The fine-scale, heterogeneous composition of urban environments results in complex spatial processes that are becoming progressively better understood with the aid of advanced RS technologies and GIS. There is an opinion that, in spite of the impressive advancement of geospatial technologies and their growing integration in recent years, the opportunity to fill the gaps in natural resources information and management has not been fully realized in the developing countries. Although substantial efforts are made in this direction, more is required in all aspects including training to produce skilled and dedicated manpower to utilize the technologies for natural resources management in urban areas.

URBAN BIODIVERSITY AND SUSTAINABILITY

According to UN-Habitat (2006), more than half the world's population is living in cities and thus challenges the conservation and sustainable use of biodiversity in urban settings and to the functioning of ecosystem services (Kohsaka 2010). The urban forest, which includes vegetation along urban streets and in urban parks, woodlots, abandoned sites, and residential areas, can comprise a significant percentage of a nation's tree canopy and contain a relatively high level of biodiversity (Alvey 2006). Therefore, it is important to establish that urban and suburban areas can be biologically

Table 1 *Overview of some recent studies related to urban biodiversity and management*

Issues addressed	*Approach/methodology*	*Conclusions/recommendations*	*References*
Biodiversity indicators for cities	'Drivers-Pressure-State-Impact-Response (DPSIR) model for development of urban and biodiversity indicators	• Linking scientific findings of recological trends with responding policies in the urban context is critical to the conservation and sustainable use of biodiversity.	Kohsaka (2010)
Geospatial tools such as geographic information systems (GIS), global positioning systems (GPS), and remote sensing in addressing urban forest management questions	Literature review	• Geospatial tools have a wide variety of applications from the creation of basic maps of structure, composition, and extent of the forest to complex forest disturbance models. • GIS and statistical software can be used to develop predictive models to delineate and monitor forest areas that are more likely to be affected in the future.	Ward and Johnson (2007)
An approach to urban forest planning and management for facilitating sustainable urban forest management	Developing a dynamic planning and management model	• Social, political, and biological concerns must be jointly addressed to sustain urban forest health and structure in the 21st century.	Dwyer, Nowak, and Noble (2003)
Diversity and density of bird and woody species in urban green spaces	Belt transects and Principal Component Analysis (PCA)	• Maintaining larger green spaces with high structural diversity may be effective in maintaining plant and bird diversity.	Khera, Mehta, and Sabata (2009)

Contd...

Table 1 *Contd...*

Issues addressed	*Approach/methodology*	*Conclusions/recommendations*	*References*
Application of landscape ecology in the planning and design of urban ecological systems	Patch-corridor-matrix model	• Strategies to restore ecological structure and develop its function in metropolitan areas are crucial for effective urban planning.	Aminzadeh and Khansefid (2010)
Spatial-temporal gradient analysis of urban green spaces	GIS, RS, and FRAGSTATS	• A new method for quantifying and capturing changes in green space patterns is presented. Supported by GIS and remote sensing, the method comprises quantification of local area green spaces by the 'moving window' technique (using FRAGSTATS), and a gradient analysis involving sampling from the urban centre to the fringe. • In comparison to the quantifying metrics in entire landscapes, this method more effectively links patterns and processes, and can establish an important basis for subsequent analysis of ecological and socio-economic functions of green spaces.	Kong and Nakagoshi (2006)
Influence of rapid urban growth and regional policies	UPlan-urban growth model and GIS	• Study may provide guidance for planners responsible for the future	Beardsley, Thorne, Roth, *et al.* (2009)

Contd...

Table 1 *Contd...*

Issues addressed	*Approach/methodology*	*Conclusions/recommendations*	*References*
on biological resources		footprint of human settlement in the San Joaquin Valley, and describes a methodology that could be used in other parts of the world.	
Relationship between green space types and people's satisfaction	Remote sensing imagery	• The area where people are most satisfied with the green space. Street-side trees are the most popular green type.	Pham and He (2008)
Extraction of urban green space under the shadow from an IKONOS image	Remote sensing imagery: a hierarchical classification technique combining textural features and spectral characteristics	• Hierarchical classification technique combining features and spectral characteristics can be used for the extraction of urban green spaces in deep shadow from IKONOS images and improve the precision of extraction.	Wang, Hong, Xu, *et al.* (2009)
Quantitative holistic evaluation technique for landscape aesthetic assessment	Photographs have been used for scenic beauty/ visual quality assessment	• A systematic method of helping fulfil objectivity and reliability in simulated landscape assessment is presented. There was strong indication that the respondents had different expectations of what a functional urban green space	Chen, Adimo, and Bao (2009)

Contd...

Table 1 *Contd...*

Issues addressed	*Approach/methodology*	*Conclusions/recommendations*	*References*
		should be in terms of auditory, tactile, and visual quality, and general recreational needs.	
Bird community species-richness, abundance, and composition from intra- and peri-urban areas in a medium-sized neotropical city to develop an easy-to-use method to calculate the area where urban and adjacent non-urban systems intermingle	Field studies method development	• Based on the ecological barrier effect that 'peri-urban' areas can represent for wildlife communities, at least for birds, the author suggested the use of four terms related to the geographical location of urban sites: the 'peri-urban', 'intra-urban', 'urban fringe', and 'extra-urban' areas of the city. The proposed method is robust and useful for a great variety of amoeboid-growing cities.	MacGregor-Fors (2010)
Effects of urban green space dispersal and density on quality of the environment	RS, GIS, and ecological indices	• The regions with the least green space area per capita and vegetation cover density are the most polluted and populated areas. This study broadly discussed planning priorities for urban green space development.	Faryadi and Taheri (2009)
Changes in the spatial structure of landscape	Landscape metrics approach	• Demonstrated the effectiveness of the method of multitemporal equal-distance area analysis	Solon (2009)

Contd...

Table 1 *Contd...*

Issues addressed	*Approach/methodology*	*Conclusions/recommendations*	*References*
		combined with the use of landscape metrics for identifying changes of landscape pattern in response to urbanization and other spatial processes. • The landscape metrics used in the study may find application in urban planning and in activities aimed at protecting landscape resources and biological diversity.	
Urban expansion and related factors	GIS and RS	• Oasis cities are expanding under the combined effects of urbanization, economic development, transportation, environment, resources, policy, and planning. Therefore, new methods that combined RS and GIS technologies to quantitatively assess urban morphology are suggested.	Liu, Zhang, Lei, *et al.* (2010)
Habitat diversity of flora and land-use patterns	Floristic surveys	• Demonstrated the importance of the historical factor for biodiversity. Recommendations are given for the maintenance	Zerbe, Maurer, Schmitz, *et al.* (2003)

Contd...

Table 1 *Contd...*

Issues addressed	*Approach/methodology*	*Conclusions/recommendations*	*References*
		and development of biological diversity in cities, focusing on landscape and habitat as well as species diversity. • Future research and planning of nature conservation and sustainable development of urban-industrial areas should take the biological diversity in cities fully into account.	
Species richness, abundance, and species composition of ant communities in three urban habitats (forest remnants, community gardens, vacant lots)	Field studies	• Implications for biodiversity conservation in urban ecosystems are discussed and suggested that to best conserve high species richness in urban areas, across all habitats, maintaining a variety of habitat types may be most beneficial.	Uno, Cotton, and Philpott (2010)
Potential of remote sensing tools in biodiversity research in an urban landscape	Field sampling and remote sensing	• Remote sensing approaches may provide planners and conservation biologists with an efficient and cost-effective method to study and estimate biodiversity across urban environments that range between	Bino, Levin, Darawshi, *et al.* (2008)

Contd...

Table 1 *Contd...*

Issues addressed	*Approach/methodology*	*Conclusions/recommendations*	*References*
		densely built-up areas, residential neighbourhoods, urban parks, and the peri-urban environment.	
Ecosystem services in functional urban regions	Literature review	• The ecosystem services approach can provide an opportunity for land-use planning to develop ecologically sustainable urban regions. Currently, information on ecosystem services of urban regions is lacking and there is a need to improve the knowledge base for land-use planning.	Niemelä, Saarela, Söderman, *et al.* (2010)
Effects of diesel exhaust emissions on herbaceous species	Experimental simulation	• The study clearly demonstrated the potential for realistic levels of vehicle exhaust pollution to have direct adverse effects on urban vegetation.	Honour, Bell, Ashenden, *et al.* (2009)
Biodiversity planning and role of ecological greenways and parks in urban species conservation	GIS, RS and GAP model	• Drew connections between the rapidly expanding literature on biodiversity conservation and highlighted the connections between greenways, especially ecological greenways, and urban biodiversity conservation.	Bryant (2006)

Contd...

Table 1 *Contd...*

Issues addressed	*Approach/methodology*	*Conclusions/recommendations*	*References*
Possible motivations for urban biodiversity conservation and improving human well-being	Literature synthesis—primary and grey literature, author's own experience, and discussions with colleagues	• Highlighted the importance of specifying the motivations behind urban biodiversity conservation and argued that if the goals are unknown, progress cannot be assessed.	Dearborn and Kark (2010)
Examples of urban biodiversity research from various areas of the world and understanding the patterns and processes that affect urban biodiversity	Literature synthesis	• The potential for urban areas to harbour considerable amounts of biodiversity needs to be recognized by city planners and urban foresters so that management practices that preserve and promotes diversity that can be pursued. • Management options should focus on increasing biodiversity in all aspects of the urban forest, from street trees to urban parks and woodlots.	Alvey (2006)
Proposal of a conceptual model and a set of hypotheses that explicitly link urban pattern to human and ecosystem functions in urban ecosystems	Literature review and conceptual model development	• The implications of the conceptual model for urban planning and design are discussed. Since ecosystem responses vary with various ecological processes, this knowledge is critical to determine what processes need to be maintained to ensure that urban ecosystem services can simultaneously support both humans and other species.	Alberti and Marzluff (2004)

Contd...

Table 1 *Contd...*

Issues addressed	*Approach/methodology*	*Conclusions/recommendations*	*References*
Past and prospective land uses changes and their drivers and impacts	RS and GIS	• The loss of important habitats, including areas of high ecological significance within the urban footprint are highly likely due to increased pressures to develop land for urban use within this designated area. • Emphasized the need to critically review and potentially give more weight to the protection and management of important habitats and landscape features within urban footprints, particularly in regions whose economy is largely built on the amenities provided by the natural environment.	Bohnet and Pert (2010)

rich so that urban foresters and city planners can actively preserve and promote that diversity. Research conducted in many countries and across various spatial scales lends support to this assertion (Balmford, Moore, Brooks, *et al.* 2001; Jim and Liu 2001; Araújo 2003; Godefroid and Koedam 2003; Cornelis and Hermy 2004; Kühn, Brandl, and Klotz 2004). Just as researchers speculate that the increase in temperature, rise in sea level, and extreme/unusual weather patterns will have a dangerous impact on habitat and ecosystems, urban biodiversity, too, may undergo greater stress and such rigorous conditions may render urban forest ecosystems more vulnerable. The urban land-use patterns change rapidly in response to economic, social, and environmental forces; hence, urban forest planning and management require rapid, accurate, and systematic methods for acquiring information (Sanders 1984; Bergen, Colwell, and Sapio 2000; Dwyer and Nowak 2000).

Sustainability is related to forest health and has become a primary objective in the present-day ecosystem management (Coppin, Jonckheere, Nackaerts, *et al.* 2004). A sustainable urban forest is one that maintains biodiversity, productivity, regenerative capacity, vitality, and the potential to fulfil relevant ecological, economic, and social functions (Wiersum 1995). The key attributes with significant implications for sustainability include its diversity, connectedness, and dynamics (Dwyer and Nowak 2000). The diversity of the urban forest is a function of variations in land use, land ownerships, and management objectives, as well as variable tree species and sizes, ground cover, soil types, microclimates, wildlife, people, buildings, and infrastructure (Dwyer and Nowak, 2000). According to the Millennium Ecosystem Assessment 2006, the key challenge faced by urban areas today is climate change, which is likely to become the dominant driver of biodiversity loss by the end of this century. The Millennium Ecosystem Assessment is an international work programme designed to meet the needs of decision makers and the public for scientific information concerning the consequences of ecosystem change for human well-being and options for responding to those changes.

Presently, we are facing increasingly the challenges of an urbanized world, and it is speculated that significant urban growth will occur in developing countries. Researchers have suggested that increased ecological investigation in urban landscapes and promotion of ecologically guided biodiversity conservation in these places are urgently required to inform future conservation efforts and promote sustainable development. A growing body of opinion also indicates that biodiversity should be looked at as an asset for poverty alleviation and sustaining livelihood because it is a critical natural capital that has to be actively managed as part of the social and economic development. It is evident that urban forests and green areas can play an important role in enhancing the sustainable development of urban settlements by their multiple functions: improving the environment through reduction of pollution, mitigating climate change impact, providing livelihood to the urban poor through forest products, and enhancing the quality of urban life. Hence, the role of geospatial tools becomes crucial for achieving sustainable urban development in the 21st century.

RS–GIS APPLICATIONS IN URBAN BIODIVERSITY MANAGEMENT

The need to understand and include urban biodiversity patterns in research and management programmes is becoming vital (Bino, Levin, Darawshi, *et al.* 2008) because urban landscape is expanding rapidly and reshaping the distribution of many animal and plant species. The increasing demands in the urban planning and management sectors have called for coordinate application of RS, GIS, and spatial modelling techniques. Remote sensing systems collect vast amounts of biological and physical data, thus allowing both inventory and monitoring of the environment possible. The GIS application can be integrated with remote sensing systems to realize the full potential of both the systems. The integration of such spatial technologies with other analytical approaches is often desirable to produce better information, thereby enhancing our understanding for better management of natural resources in the urban areas.

In general, biodiversity changes have been studied across urban gradients (McDonnell and Pickett 1990; Alberti, Botsford, and Cohen 2001; Marzluff, Bowman, and Donnelly 2001; Chace and Walsh 2006) and declining diversity from the city periphery to the core is reported (McDonnell and Pickett 1990; Clergeau, Savard, Mennechez, *et al.* 1998; Cam, Nichols, Sauer, *et al.* 2000; Melles, Glenn, and Martin 2003). Field-based estimates of land cover and species richness are expensive and resource consuming, especially when carried out over large areas and at multiple spatial scales (Bino, Levin, Darawshi, *et al.* 2008). Remote sensing is easily applied and enables monitoring of large areas at low cost. It is, therefore, a useful tool for conservation scientists and for decision makers (Diamond 1988; Fairbanks and McGwire 2004). Driven by technological advancement and societal needs, remote sensing of urban areas has increasingly become the new arena of geospatial technology and finds applications in all socioeconomic sectors as well (Weng and Quattrochi 2006). Recent advancement in the technology of remote sensing has brought about miracles in the availability of high-resolution satellite imagery (Cartosat I and II, IKONOS, Quickbird imagery). This technological advancement and increasing availability of remotely sensed observation facilitates the development of new tools and approaches for understanding urban environment (Ben-Dor, Levin, and Saaroni 2001; Miller and Small 2003). The details about remote sensing platforms and sensor application in urban studies are presented in Table 2.

Recent studies have shown that RS and GIS tools can be useful in examining biodiversity patterns in urban landscapes, land-use and land-cover classification, change detection, and monitoring urban green spaces, clearly demonstrating the importance of integrated satellite remote sensing images in conjunction with a GIS database. To facilitate RS–GIS-based research in the urban environment, high-resolution satellite images and most land-use and planning maps are available in digital format. With technological advancement, correlating various layers of information about a feature from satellite imagery is feasible with the help of image processing software such as ERDAS Imagine, ENVI, PCI Geomatica, and ILWIS. The resultant superimposed map in GIS software (Map info, ArcView, ArcGIS) can provide valuable information for planning, implementation and management of urban areas. Besides, several software packages are open source (Table 3) and can be used by researchers to address

Table 2 *Common remote sensing satellite sensors used in urban studies*

Satellite and sensor system	*Spatial resolute (meter, pixel)*	*Year of operation*	*Mapping scale*	*Extractable information*
Landsat (MSS)	80	1972	1:1 000 000	Broad land-use/land-cover and urban
IRS-1A and 1B (LISS-I)	72	1988 and 1991	1:250 000	sprawl
Landsat TM	30	1982	1:50 0000	Thematic data for broad structural
IRS-1A and 1B (LISS-II)	36	1988 and 1991		plans and spatial strategies
IRS-1C and 1D (LISS-III)	23	1995 and 1997		
SPOT HRV-I (MLA)	20	1998		
IRS-1D (LISS-IV)	5.8	2003	1:5 000	
ASTER			1:250 000	Land-use/land-cover, urban sprawl,
VNIR (0.52–0.86 μm)	15	1999	1:50 000	ecological monitoring data
SWIR (1.60–2.43 μm)	30			
TIR (8.125–11.65 μm)	90			
SPOT HRV-II (MLA)	10	1998	1:25 000	Data for land-use/land-cover for
IRS-1C and 1-D (PAN)	5.8	1995 and 1997	1:10 000	urban area
MOMS-II	4	1983	1:8 000	Land-use/land-cover details
IKONOS	1.0	1999	1:4 000	Cadastral map, detailed information
Quickbird	0.61	2001	1:2 000	extraction for urban planning, and infrastructure mapping
CARTOSAT-1	2.5	2005	1:4 000	Large-scale cartographic work and
CARTOSAT-2	1.0	2007	1:1 000	DM generation cartographic
			1:2 000	applications at cadastral level, urban and rural infrastructure development and management
RESOURCESAT-I	5.8	2003	1:10 000	Monitoring the urban growth,
(LISS-IV)			1:4 000	inventory of land-use/ land-cover

Source Adapted from Verma, Kumari, and Tiwary (2009)

various research issues pertaining to urban biodiversity monitoring, management, and sustainable development planning.

FUTURE RESEARCH

At present, research issues such as biodiversity, climate change, and sustainable development are a priority among ecologists and conservation biologists. Habitat loss, habitat fragmentation, and habitat degradation are the greatest threats to biodiversity and are the three main processes contributing to landscape change (Fischer and Lindenmayer 2007). Some of the changes as a result of climate change include the following.

- Accelerated habitat loss: Specifically, small islands of habitats may not be able to sustain themselves as the climate changes.
- Changes in range: The geographical range of many species will move towards the poles or higher altitudes.
- Increased extinction rates: Many species that are threatened are likely to become extinct.

To formulate comprehensive management plans in order to sustain urban biodiversity, many researchers highlight the need for long-term monitoring programs focusing on how and why urban biodiversity changes through time and across urbanization gradients. Also, the combination of economic, socio-political, and biophysical factors is considered to have substantial effects on urban ecosystems integrity. Thus, a key component of urban research should focus on developing integrated approaches that define the resilience of urban ecosystems to these forces of change and maintain the consistent flow of ecosystem goods and services. In this context, RS and GIS techniques could be utilized in the identification of sensitive and vulnerable habitats, development of a strategy to protect and restore urban vegetation for conservation, and guidance for sustainable urban development.

Some other important issues for urban research include the following.

- Understanding changes in biodiversity in response to urbanization, and investigating the ecological consequences of urban growth.

Table 3 *Some important open source/non-commercial GIS products*

Product name/type	*Supplier, web link*	*Features*
GRASS/GIS	Open source, <www.grass.itc.it>	Open source GIS with both raster and vector support. Geographic Resources Analysis Support System (GRASS) is a Geographic Information System (GIS) used for geospatial data management and analysis, image processing, graphics/maps production, spatial modeling, and visualization. GRASS is currently used in academic and commercial settings around the world, as well as by many governmental agencies and environmental consulting companies.
ILWIS	ITC, <www.itc.nl/ilwis>	The Integrated Land and Water Information System (ILWIS) is a PC-based GIS and remote sensing (RS) software developed by ITC. ILWIS comprises a complete package of image processing, spatial analysis, and digital mapping.
GeoDa/Exploratory data analysis	Spatial Analysis Laboratory, University of Connecticut, USA, <www.sal.uiuc.edu>	Exploratory spatial data analysis, vector (L Anselin). GeoDa is the latest incarnation of a collection of software tools designed to implement techniques for exploratory spatial data analysis (ESDA) on lattice data. It is intended to provide a user-friendly and graphical interface to methods of descriptive spatial data analysis such as global and local (LISA) autocorrelation statistics and indicators of spatial outliers, plus some more advanced regression analysis facilities.
PASSaGE/Specialized data analysis	M Rosenberg, Arizona State University, <www.passagesoftware.net>	Pattern Analysis, Spatial Statistics, and Geographic Exegesis (PASSaGE) is an easy-to-use program for general spatial analysis and includes a wide range of analysis functions. It supports multiple OS (Win, Mac, Unix).

Contd...

Table 3 *Contd...*

Product name/type	*Supplier, web link*	*Features*
SAM/Statistical analysis	Instituto de CiênciasBiológicas Universidade Federal de Goiás, Brazil, <www.ecoevol.ufg.br/sam>	Spatial Analysis in Macroecology (SAM) is a compact but robust computer program designed as a package of statistical tools for spatial analysis, mainly for applications in macroecology, geographical ecology, and biogeography. SAM offers a wide spectrum of statistical methods used in Surface Pattern Spatial Analysis, including a range of basic stats, autocorrelation analyses, and regression models (GLIM, SAR, CAR, GWR).
SAGA/GIS	Abteilungfür Physische Geographie Geographisches Institut Göttingen, Germany, <www.saga-gis.org>	Open source GIS designed with geosciences in mind, especially terrain and hydrographic analysis. Powerful raster analysis and programmability
Landserf/Terrain analysis	Jo Wood, Department of Information Science, City University, London, UK, <www.landserf.org>	Java-based surface analysis package, cross-platform, with excellent analysis and visualization facilities
OSSIM/Image handling	Open source, <www.ossim.org>	Open Source Software Image Map (OSSIM) is a high performance engine for remote sensing, image processing, geographical information systems, and photogrammetry.
QGIS/GIS	Open source, <www.qgis.org>	Quantum GIS (QGIS) is a user-friendly open source GIS that runs on Linux, Unix, Mac OSX, and Windows. QGIS supports vector, raster, and database formats.
Fragstats/Landscape analysis	University of Massachusetts, USA, <www.umass.edu/landeco/research/fragstats/fragstats.html>	Analysis of ecological raster data. Spatial pattern analysis for categorical maps.
TNTMips/GIS	Microimages, Lincoln, NA, USA, <www.microimages.com>	Commercial, generic, cross-platform GIS developed from image processing background. Extensive analytics toolset.

Contd...

Table 3 *Contd...*

Product name/type	*Supplier, web link*	*Features*
STARS/Spatio-temporal analysis	Regional Analysis Lab, San Diego University, CA, USA, <www.regionalanalysislab.org/index.php/Main/STARS>	Space-time analysis of regional systems (STARS) is an open source environment written in Python. It supports exploratory dynamic spatial data analysis. STARS contain a set of exploratory spatial data analysis modules, together with several newly developed measures for space-time analysis.

Source *<www.spatialanalysisonline.com/softwareFree.pdf>*

- Analysing changes in the spatial characteristics of urban vegetation with the combined use of remote sensing, GIS, and spatial metrics.
- Assessing the impact of climate on biodiversity and addressing the issues of urban heat islands, urban metabolism, and carrying capacity. This aspect is very relevant because current climate changes coupled with human pressures are stressing biodiversity far beyond the levels imposed by the global climatic change that occurred in the recent evolutionary past.
- Working on benchmark studies covering a wide range of ecosystems by focusing on ensuring long-term sustainability of ecosystems and the critical services they provide to humanity.
- Investigating how climate change or habitat fragmentation would affect the population dynamics of important species.

More importantly, collaboration among stakeholders, public participation, and the integration of temporal and spatial scales is critical to the conservation and sustainable use of urban biodiversity.

CONCLUSION

This review has provided an overview of the usefulness of geospatial tools with particular emphasis on RS and GIS techniques for monitoring urban biodiversity and sustainable management of natural resources. The intelligent use of these technologies could contribute a great deal to this emerging research area because understanding the status of urban biodiversity and habitat fragmentation by land-use changes and its causes are important in decision-making and in creating public awareness about conserving natural resources of urban areas. From the point of view of biodiversity studies, urban ecosystems need to receive utmost attention for the goods and services they provide to urban dwellers under changing environment. The conservation of biological diversity and ecosystem services in urban environments provide valuable ecological information, which could be incorporated into resources planning and management. In this context, the RS–GIS techniques, coupled with various modern modelling approaches, become crucial. These tools can effectively help in addressing the issues of biodiversity loss, assessing ecosystem integrity, quantifying the provision of goods and services provided by urban biodiversity, minimizing various threats to biodiversity, and promoting sustainable biodiversity management in urban landscapes. Such information has great relevance in scenario development for future urbanization, possible environmental impacts, and sustainable development of urban landscapes. In addition, the RS–GIS tools can also help in understanding how climate and land-use change will affect plant communities and the ecosystem function, and predict future biodiversity patterns, which are critical for the future management of urban ecosystems in a sustainable manner.

ACKNOWLEDGEMENTS

The authors would like to thank the Universiti Sains Malaysia for its facilities and support. Sanjay Gairola is especially grateful to the Universiti Sains Malaysia for the award of postdoctoral fellowship.

REFERENCES

Alberti, M., E. Botsford, and A. Cohen. 2001. Quantifying the urban gradient: linking urban planning and ecology. In *Avian Ecology and Conservation in an Urbanizing World*, edited by J. M. Marzluff, R. Bowman, and R. Donnelly, pp. 89–115. Massachusetts: Kluwer Academic Publishers

Alberti, M., and J. M. Marzluff. 2004. Ecological resilience in urban ecosystems: linking urban patterns to human and ecological functions. *Urban Ecosystems* 7:241–265

Alberti, M., J. M. Marzluff, E. Shulenberger, G. R. C. Bradley, and C. Zumberunnen. 2003. Integrating humans into ecology: opportunities and challenges for studying urban ecosystems. *Journal of Bioscience* 53(12):1169–1179

Alvey, A. A. 2006. Promoting and preserving biodiversity in the urban forest. *Urban Forestry and Urban Greening* 5:195–201

Aminzadeh, B., and M. Khansefid. 2010. Improving the natural and built ecological systems in an urban environment. *International Journal of Environmental Research* 4(2):199–208

Araújo, M. B. 2003. The coincidence of people and biodiversity in Europe. *Global Ecology and Biogeography* 12:5–12

Balmford, A., J. L. Moore, T. Brooks, N. Burgess, L. A. Hansen, P. Williams, and C. Rahbek. 2001. Conservation conflicts across Africa. *Science* 291:2616–2619

Bauer, M., F. Yuan, and K. Sawaya. 2003. Multi-temporal Landsat image classification and change analysis of land cover in the Twin Cities (Minnesota) metropolitan area. [Muti-Temp-2003, Second International Workshop on the Analysis of Multitemporal Remote Sensing Images, 16–18 July, Ispra, Italy]

Beardsley, K., J. H. Thorne, N. E. Roth, S. Gao, and M. C. McCoy. 2009. Assessing the influence of rapid urban growth and regional policies on biological resources. *Landscape and Urban Planning* 93:172–183

Ben-Dor, E., N. Levin, and H. Saaroni. 2001. A spectral based recognition of the urban environment using the visible and near-infrared spectral region (0.4–1.1 nm): a case study over Tel-Aviv. *International Journal of Remote Sensing* 22(11):2193–2218

Bergen, K., J. Colwell, and F. Sapio. 2000. Remote sensing and forestry: collaborative implementation for a new century of forest information solutions. *Journal of Forestry* 98:4–9

Bino, G., N. Levin, S. Darawshi, N. Van der Hal, A. Reich-Solomon, and S. Kark. 2008. Accurate prediction of bird species richness patterns in an urban environment using Landsat-derived NDVI and spectral unmixing. *International Journal of Remote Sensing* 29(13):3675–3700

Bohnet, I. C., and P. L. Pert. 2010. Patterns, drivers and impacts of urban growth: a study from Cairns, Queensland, Australia, from 1952 to 2031. *Landscape and Urban Planning* 97:239–248

Booth, D., and C. Jackson. 1997. Urbanization of aquatic systems: degradation thresholds, stormwater detention, and the limits of mitigation. *Journal of the American Water Resources Association* 33:1077–1090

Bryant, M. M. 2006. Urban landscape conservation and the role of ecological greenways at local and metropolitan scales. *Landscape and Urban Planning* 76:23–44

Cam, E., J. D. Nichols, J. R. Sauer, J. E. Hines, and C. H. Flather. 2000. Relative species richness and community completeness: birds and urbanization in the Mid-Atlantic states. *Ecological Applications* 10:1196–1210

Chace, J. F., and J. J. Walsh. 2006. Urban effects on native avifauna: a review. *Landscape and Urban Planning* 74:46–69

Chen, B., O. A. Adimo, and Z. Bao. 2009. Assessment of aesthetic quality and multiple functions of urban green space from the users' perspective:

the case of Hangzhou Flower Garden, China. *Landscape and Urban Planning* 93:76–82

Chiesura, A. 2004. The role of urban parks for the sustainable city. *Landscape and Urban Planning* 68:129–138

Clergeau, P., J. P. L. Savard, G. Mennechez, and G. Falardeau. 1998. Bird abundance and diversity along an urban–rural gradient: a comparative study between two cities on different continents. *Condor* 100:413–425

Colding, J. 2007. Ecological land-use complementation for building resilience in urban ecosystems. *Landscape and Urban Planning* 81:46–55

Coppin, P., I. Jonckheere, K. Nackaerts, B. Muys, and E. Lambin. 2004. Digital change detection methods in ecosystem monitoring: a review. *International Journal of Remote Sensing* 25:1565–1596

Cornelis, J., and M. Hermy. 2004. Biodiversity relationships in urban and suburban parks in Flanders. *Landscape and Urban Planning* 69:385–401

Dearborn, D. C., and S. Kark. 2010. Motivations for conserving urban biodiversity. *Conservation Biology* 24(2):432–440

Diamond, J. M. 1988. Conservation biology: red books or green lists. *Nature* 332:304–305

Dragicevic, S. 2008. GeoComputation: modeling with spatial agents. *Computers, Environment and Urban Systems* 32:415–416

Dwyer, J. F., and D. J. Nowak. 2000. A national assessment of the urban forest: an overview. In *Proceedings of the Society of American Foresters 1999 National Convention*, Portland, Oregon, 11–15 September 1999, pp. 157–162. Maryland: Society of American Foresters

Dwyer, J. F., D. J. Nowak, and M. H. Noble. 2003. Sustaining urban forests. *Journal of Arboriculture* 29(1):49–55

Ellis, E. C., and N. Ramankutty. 2008. Putting people in the map: anthropogenic biomes of the world. *Frontiers in Ecology and the Environment* 6:439–447

Fairbanks, D. H. K., and K. C. Mcgwire. 2004. Patterns of floristic richness in vegetation communities of California: regional scale analysis with multi-temporal NDVI. *Global Ecology and Biogeography* 13:221–235

Faryadi, Sh., and Sh. Taheri. 2009. Interconnections of urban green spaces and environmental quality of Tehran. *International Journal of Environmental Research* 3(2):199–208

Fischer, J., and D. B. Lindenmayer. 2007. Landscape modification and habitat fragmentation: a synthesis. *Global Ecology and Biogeography* 16:265–280

Godefroid, S., and N. Koedam. 2003. Distribution pattern of the flora in a peri-urban forest: an effect of the city-forest ecotone. *Landscape and Urban Planning* 65:169–185

Grimm, N. B., S. H. Faeth, N. E. Golubiewski, C.L. Redman, J. Wu, X. Bai, and J. M. Briggs. 2008. Global change and the ecology of cities. *Science* 319:756–760

Grimmond, C. S. B., and C. Souch. 1994. Surface description for urban climate studies: a GIS based methodology. *Geocarto International* 9:47–59

Hansen, A. J., R. L. Knight, J. M. Marzluff, S. Powell, K. Brown, and P. H. Gude. 2005. Effects of exurban development on biodiversity: patterns, mechanisms, and research needs. *Ecological Applications* 15(6):1893–1905

Herold, M., N. C. Goldstein, and K. C. Clarke. 2003. The spatiotemporal form of urban growth: measurement, analysis and modeling. *Remote Sensing of Environment* 86(3):286–302

Hobbs, R. 1997. Future landscapes and the future of landscape ecology. *Landscape and Urban Planning* 37:1–9

Honour, S. L., J. N. B. Bell, T. W. Ashenden, J. N. Cape, and S. A. Power. 2009. Responses of herbaceous plants to urban air pollution: effects on growth, phenology and leaf surface characteristics. *Environmental Pollution* 157:1279–1286

Jensen, J., and J. Im. 2007. Remote sensing change detection in urban environments. In *Geo-spatial Technologies in Urban Environments Policy, Practice, and Pixels*, edited by R. R. Jensen, J. D. Gatrell, and D. D. McLean, pp. 7–32. Berlin: Springer-Verlag

Jim, C. Y., and S. S. Chen. 2003. Comprehensive green space planning based on landscape ecology principles in compact Nanjing city, China. *Landscape and Urban Planning* 65:95–116

Jim, C. Y., and W. Y. Chen. 2006a. Perception and attitude of residents towards urban green spaces in Guangzhou (China). *Environmental Management* 38(3):338–349

Jim, C. Y., and W. Y. Chen. 2006b. Recreation-amenity use and contingent valuation of urban green spaces in Guangzhou, China. *Landscape and Urban Planning* 75(1, 2):81–96

Jim, C. Y., and H. T. Liu. 2001. Species diversity of three major urban forest types in Guangzhou City, China. *Forest Ecology and Management* 146:99–114

Johnson, C. P., B. Deshmukh, and M. Kale. 2008. Role of GIS and remote sensing in the sustainable development of Mauritius. In *Proceedings of the Tenth International Conference for Spatial Data Infrastructure (GSDI-10): Small Island Perspectives on Global Challenges: The Role of Spatial Data in Supporting a Sustainable Future*, 25–29 February 2008, St. Augustine, Trinidad

Kazmierczak, A. E., and P. James. 2008. Planning for biodiversity conservation in large urban areas: the ecological framework for Greater Manchester. *Salzburger Geographische Arbeiten* 42:129–149

Khera, N., V. Mehta, and B. C. Sabata. 2009. Interrelationship of bird and habitat features in urban green spaces in Delhi, India. *Urban Forestry and Urban Greening* 8:187–196

Kinzig, A. P., P. Warren, C. Martin, D. Hope, and M. Katti. 2005. The effects of human socioeconomic status and cultural characteristics on urban patterns of biodiversity. *Ecology and Society*. Details available at <www.ecologyandsociety.org/vol10/iss1/art23>

Kohsaka, R. 2010. Developing biodiversity indicators for cities: applying the DPSIR model to Nagoya and integrating social and ecological aspects. *Ecological Research* 25: 925–936

Kong, F., and N. Nakagoshi. 2006. Spatial-temporal gradient analysis of urban green spaces in Jinan, China. *Landscape and Urban Planning* 78:147–164

Kühn, I., R. Brandl, and S. Klotz. 2004. The flora of German cities is naturally species rich. *Evolutionary Ecology Research* 6:749–764

Liu, Y., X. Zhang, J. Lei, and L. Zhu. 2010. Urban expansion of oasis cities between 1990 and 2007 in Xinjiang, China. *International Journal of Sustainable Development and World Ecology* 17(3):253–262

Longley, P. A. 2002. Geographical information systems: will developments in urban remote sensing and GIS lead to 'better' urban geography? *Progress in Human Geography* 26(2):231–239

Luck, M., and J. Wu. 2002. A gradient analysis of urban landscape pattern: a case study from the Phoenix metropolitan region, Arizona, USA. *Landscape Ecology* 17:327–339

Lyons, J. R. 1997. Urban ecosystem management: bringing science and policy together. *Urban Ecosystems* 1:77–83

MacGregor-Fors, I. 2010. How to measure the urban-wildland ecotone: redefining 'peri-urban' areas. *Ecological Research* 25:883–887

Marzluff, J. M., R. Bowman, and R. E. Donnelly. 2001. A historical perspective on urban bird research: trends, terms, and approaches. In

Avian Conservation and Ecology in an Urbanizing World, edited by J. M. Marzluff, R. Bowman, and R. E. Donnelly, pp. 1–17. Norwell: Kluwer Academic

Millenium Ecosystem Assessment. 2005. *Ecosystems and Human Well-being: Biodiversity Synthesis*. Washington, DC: World Resources Institute

Mcdonnell, M. J., and S. T. A. Pickett. 1990. Ecosystem structure and function along urban rural gradients: an unexploited opportunity for ecology. *Ecology* 71:1232–1237

McIntyre, N. E. 2000. Ecology of urban arthropods: a review and a call to action. *Annals of the Entomology Society of America* 93:825–835

McIntyre, N. E., J. Rango, W. F. Fagan, and S. H. Faeth. 2001. Ground arthropod community structure in a heterogeneous urban environment. *Landscape and Urban Planning* 52:257–274

McKinney, M. L. 2002. Urbanization, biodiversity, and conservation. *BioScience* 52:883–890

McKinney, M. L. 2008. Effects of urbanization on species richness: a review of plants and animals. *Urban Ecosystems* 11:161–176

Melles, S., S. Glenn, and K. Martin. 2003. Urban bird diversity and landscape complexity: species–environment associations along a multiscale habitat gradient. *Conservation Ecology* 7(1):5

Miller, J. R., S. A. Snyder, A. M. Skibbe, and R. G. Haight. 2009. Prioritizing conservation targets in a rapidly urbanizing landscape. *Landscape and Urban Planning* 93:123–131

Miller, J. R., and R. J. Hobbs. 2002. Conservation where people live and work. *Conservation Biology* 16(2):330–337

Miller, R. S., and C. Small. 2003. Cities from space: potential applications of remote sensing in urban environmental research and policy. *Environmental Science and Policy* 6:129–137

Niemelä, J., S. R. Saarela, T. Söderman, L. Kopperoinen, V. Yli-Pelkonen, S. Väre, and D. J. Kotze. 2010. Using the ecosystem services approach for better planning and conservation of urban green spaces: a Finland case study. *Biodiversity and Conservation* 19:3225–3243

Olaf, B. 2000. Landscape classification in Saxony (Germany): a tool for holistic regional planning. *Landscape and Urban Planning* 50:145–155

Pham, T. T. H., and D. C. He. 2008. How do people perceive the city's green space? A view from satellite imagery (In Hanoi, Vietnam). In

Proceedings of the 2008 IEEE International Geoscience and Remote Sensing Symposium, 6–11 July 2008, III-1228-III-1231, Boston, MA, The Institute of Electrical and Electronics Engineers Inc. Details available at <www.ieeexplore.ieee.org/stamp/stamp.jsp?tp=&arnumber=4779579&isnumber=4779256>

Rafiee, R., A. S. Mahiny, and N. Khorasani. 2009. Assessment of changes in urban green spaces of Mashad city using satellite data. *International Journal of Applied Earth Observation and Geoinformation* 11:431–438

Rango, J. J. 2005. Arthropod communities on creosote bush (*Larrea tridentate*) in desert patches of varying degrees of urbanization. *Biodiversity and Conservation* 14:2185–2206

Sadler, J. P., E. C. Small, H. Fiszpan, M. G. Telfer, and J. Niemelä. 2006. Investigating environmental variation and landscape characteristics of an urban–rural gradient using woodland carabid assemblages. *Journal of Biogeography* 33:1126–1138

Sanders, R. A. 1984. Some determinants of urban forest structure. *Urban Ecology* 8:13–27

Shochat, E., P. S. Warren, S. H. Faeth, N. E. Mcintyre, and D. Hope. 2006. From patterns to emerging processes in mechanistic urban ecology. *Trends in Ecology and Evolution* 21:186–191

Smith, R. M., P. H. Warren, K. Thompson, and K. J. Gaston. 2006. Urban domestic gardens (VI): environmental correlates of invertebrate species richness. *Biodiversity and Conservation* 15:2415–2438

Solon, J. 2009. Spatial context of urbanization: landscape pattern and changes between 1950 and 1990 in the Warsaw metropolitan area, Poland. *Landscape and Urban Planning* 93:250–261

Tang, J., Y. Xue, T. Yuc, and Y. N. Guan. 2005. Aerosol optical thickness determination by exploiting the synergy of TERRA and AQUA MODIS. *Remote Sensing of Environment* 94:327–334

Tang, Z., B. A. Engel, B. C. Pijanowski, and K. J. Lim. 2005. Forecasting land use change and its environmental impact at a watershed scale. *Journal of Environmental Management* 76:35–45

Terkenli, T. S. 2001. Towards a theory of landscape: the Aegean landscape as a cultural image. *Landscape and Urban Planning* 57:239–254

Theobald, D. M., N. T. Hobbs, T. Bearly, J. A. Zack, T. Shenk, and W. E. Riebsame. 2000. Incorporating biological information in local land-use decision making: designing a system for conservation planning. *Landscape Ecology* 15:35–45

Thompson, B., and S. McLachlan. 2007. The effects of urbanization on ant communities and myrmecochory in Manitoba, Canada. *Urban Ecosystems* 10:43–52

Turner, W. R., T. Nakamura, and M. Dinetti. 2004. Global urbanization and the separation of humans from nature. *Bioscience* 54:585–590

Tzoulas, K., K. Korpela, S. Venn, V. Yli-Pelkonen, A. Kazmierszak, and J. Niemela. 2007. Promoting ecosystem and human health in urban areas using green infrastructure: a literature review. *Landscape and Urban Planning* 81:167–178

UN-Habitat 2006. *The State of the World's Cities 2006/07*. Oxford, UK: Earthscan

Uno, S., J. Cotton, and S. M. Philpott. 2010. Diversity, abundance, and species composition of ants in urban green spaces. *Urban Ecosystems* 13:425–441

Verma, R. K., S. Kumari, and R. K. Tiwary. 2009. Application of remote sensing and GIS technique for efficient urban planning in India. In *Proceedings of the Geomatrix Conference* (Geomatrix '09) 28 February–1 March 2009. CSRE, IIT Bombay, India. Details available at <www.csre.iitb.ac.in/~csre/conf/wp-content/uploads/fullpapers/OS4/OS4_13.pdf>

Wang, J., Z. Hong, S. Xu, and P. Xiao. 2009. Extraction of urban green space in shadow area from IKONOS image. 2009. Urban Remote Sensing Joint Event. Details available at <www.ieeexplore.ieee.org/xpls/abs_all.jsp?arnumber=5137698>

Ward, K. T., and G. R. Johnson. 2007. Geospatial methods provide timely and comprehensive urban forest information. *Urban Forestry and Urban Greening* 6:15–22

Weng, Q., and D. A. Quattrochi. 2006. Thermal remote sensing in urban areas: an introduction to the special issue. *Remote Sensing of Environment* 104:119–122

Wiersum, K. F. 1995. 200 years of sustainability in forestry: lessons from history. *Environmental Management* 19:321–329

Wilkie, D. S., and J. T. Finn. 1996. *Remote Sensing Imagery for Natural Resources Monitoring: a guide for first-time users.* New York: Columbia University Press

Wittig, R. 1998. Urban development and the integration of nature: reality or fiction? In *Urban Ecology*, edited by J. Breuste, H. Feldmann, and O. Uhlmann, pp. 593–599. Berlin: Springer

Yamaguchi, T. 2004. Influence of urbanization on ant distribution in parks of Tokyo and Chiba City, Japan. *Ecological Research* 19:209–216

Yuan, F., K. E. Sawaya, C. B. Leoffelholz, and M. E. Bauer. 2005. Land cover classification and change analysis of the Twin Cities (Minnesota) metropolitan area by multitemporal Landsat remote sensing. *Remote Sensing of Environment* 98:317–328

Zerbe, S., U. Maurer, S. Schmitz, and H. Sukopp. 2003. Biodiversity in Berlin and its potential for nature conservation. *Landscape and Urban Planning* 62:139–148

WEBSITES

<www.eijournal.com>

<www.geog.ucsb.edu/~jeff/115a/remotesensinghistory.html>

<www.worldcities.com.sg/speech4.php>

<www.geospatialworld.net>

<www.countdown2010.net>

<www.cbd.int/doc/press/2010/pr-2010-10-29-cop-10-en.pdf>

<www.fh-erfurt.de/urbio/httpdocs/content/keynote_speakers.php>

<www.spatialanalysisonline.com/softwareFree.pdf>

Section II

Community, Livelihood, and Management of Biodiversity

CHAPTER 9

From Cabbage to Conservation: Case of Community Forest Conservation by Hmong Tribe in Northern Thailand

Chandra Shekhar Silori

INTRODUCTION

The results presented in this chapter are based on a case study of community-initiated forest conservation in one of the well-known protected areas in Northern Thailand known as Doi Suthep-Pui National Park (DSNP). Delineated in 1981, DSNP covers an area of 261 km^2. It is located very close to Chiang Mai, one of the largest cities in Northern Thailand. Like many other forest areas in Northern Thailand, the forests in and around DSNP also reported heavy deforestation in the past (Renard 1997). Published literature has reported the Hmong tribe, one of the several ethnic groups in Northern Thailand and surrounding countries, and collectively known as hill tribes, as the main agents to convert forested areas into agricultural cultivations through shifting cultivation practices. Specifically, in DSNP large-scale deforestation during the 1970s and 1980s resulted in the loss of about 17% of the park area (Thailand Development Research Foundation 1997). This caused a serious negative impact on environmental services, mainly the hydrological system. Drying up of the major streams resulted in severe water shortage for drinking and agricultural purposes.

The inhabitants of Ban Mae Sa Mai (BMSM) village, located inside DSNP, were affected the most by these negative changes. The local residents of the village, the Hmong people, decided to address the problem themselves, since it was directly related to the security of their livelihood. They moved down from nearly 1300 m elevation to its present location at 1000 m elevation during the 1960s and initiated the process of reversing the impact of deforestation through community conservation and regeneration of degraded areas in the upper slopes inside DSNP. This case study in BMSM village thus focuses on understanding the process of community initiatives for forest recovery and the consequent environmental and social benefits. This was one of the four case studies conducted by the author, as part of Asia Fellowship programme, focusing on understanding the links between biodiversity conservation and poverty alleviation in Thailand (Silori 2010).

STUDY AREA

Ban Mae Sa Mai is located inside DSNP in the Mae Rim district of Chiang Mai province in Northern Thailand. It is one of the largest Hmong hill tribe settlements in Northern Thailand, established during the mid 1960s. DSNP is part of the greater Thanon Thongchai mountain ranges. The altitudinal variation within the park ranges from 330 m to 1685 m above mean sea level. Owing to its mountainous and varied topography, the park also acts as a watershed for several rivers, streams, and basins providing water for the city of Chiang Mai. DSNP has a diversity of forest ecosystems and habitats, represented by diptocarpus, mixed deciduous, dry, and moist evergreen and pine forests. The floral diversity is represented by over 2000 plant species, while the animal diversity is largely represented by more than 300 bird species, nearly 30 mammal species (mainly small), and a number of butterfly, insect, and reptile species (Lieorungruang and Muangyai 2008).

METHODOLOGY

Field Data Collection

With the growing population, BMSM was bifurcated around 2003/04 into two hamlets; the original BMSM and the new hamlet named as Ban Mae Sa Noi (BMSN). The village (Ban) named after the Mae Sa stream makes a boundary between the two settlements. Both the hamlets together were inhabited by a total of 216 households and a human population of nearly 1860. The field study mainly focused on the original village, BMSM, inhabited by 130 households. For field data collection, the selection of samples followed stratified random sampling methods, collecting data from a total of 17 households (13%) of the village. For field data collection, participatory research methods such as semi-structured questionnaire interviews, focused group discussions, and key informant interviews were employed. The field data collection, mainly translation with the respondents, was supported by the students from the Forest Restoration Research Unit of Chiang Mai University (FORRU-CMU). However, before collecting participatory interviews, a couple of exploratory visits were made to the study village for rapid appraisals of the general socio-economic situation and forest conservation practices with the help of the key informants. This preliminary visit and a review of the published literature (for example, Delang 2002; FORRU 2006; Elliot and Kuaraksa 2008) helped in designing semi-structured questionnaires for field surveys with the respondents.

The focus of the participatory questionnaire interview with the respondents was mainly to understand the general socio-economic status, income patterns and sources, and dependence on the forest for various purposes at the household level. Besides, qualitative information was also generated recording the perception of the respondents on how community-initiated forest restoration helped in conserving forest biodiversity and provided livelihood support. The roles of different stakeholders and the relationship of the Hmong with these stakeholders, particularly the park authorities, was also recorded to understand the capacity of the local communities to generate support from different stakeholders and ensure sustainability of their efforts.

RESULTS

Demographic and Socio-economic Scenario

The Hmong originally came from the basin of the Yangtze River and were driven south by the expanding Han Chinese population (Mottin 1980). They settled in the Himalayan highlands and became one of the 100 or more minority groups living in that area. The history of the Hmong became intertwined with global political economy when they got involved in the opium production after China lost the second opium war and encouraged the minorities living in the Himalayan highlands to grow opium (Renard 1997). Opium grows best at 1000 m above mean sea level, where cultivable land is relatively scarce. Because of this and the fact that opium can only be grown for up to 15–20 years on the same land, the Hmong were in regular migration. This brought them from China to Vietnam, Laos, and Thailand, where they continued growing opium (Geddes 1976).

In Thailand, opium cultivation was outlawed in 1959 (Renard 1997), but the Hmong continued to grow it, albeit at a decreasing rate until the late 1980s. Of late, the Hmong have a settled lifestyle and the practices of slash-and-burn have reduced significantly. Particularly in the case of BMSM, swidden cultivation was stopped long ago and local inhabitants started settled cultivation dominated by cash crops such as cabbage, vegetables, fruits, and a few traditional crops, including upland rice and maize. Over the years, there have been changes in the agricultural practices and the earlier dominated cash crops such as cabbage are now replaced by lychee plantations on the hill slopes. In the study village, almost all the respondents reported having cultivated lands, though none of them had tenurial rights over them. The cultivated area ranged from 4 rai to 25 rai (6.25 rai = 1 ha), averaging around 13 rai (2.08 ha) per household. Marginal and small farmers still grow corn, upland rice, and vegetables such as cabbage, bottle gourd, potato, lettuce, and carrot. Some farmers have recently started growing coffee and fruit trees such as mango and avocado (*Persea americana*).

Next to agriculture, other major occupations included the trading of silver ornaments, precious stones, jewellery, and cloth in the weekly market in Chiang Mai, about 50 km from the villages. This involved about 47% of the total sampled households of the village. Another 29% of the households earned their additional income by working as daily wage labourers in the agricultural and construction sectors. Of the remaining households, 12% earned their subsidiary income by working in the government and/or private sector, and offering home stay services for a limited number of tourists who visited the area to enjoy nature.

Household Income Patterns

Household income ranged from 41000 to 468000 baht, averaging 186584 baht/annum for each household. The highest contribution to the average income was agriculture (62%), followed by trade (27%), and other sources such as employment (6.7%) and daily wage labour (3.7%) (Figure 1). To understand the income patterns across different economic groups, the sampled households were categorized into three broad categories based on the land area they cultivated. These included small farmers (less than 10 rai), medium farmers (10–15 rai), and large farmers (more than 15 rai). The average annual income ranged from 117991 baht for small landholders to 291540 baht for large landholders (Table 1). Across all the three landholding categories, the agricultural sector contributed substantially to the average household income, ranging from nearly 50% for large landholders to 73% for small landholders. For medium and large landholders, trade was the second major source of income, contributing nearly 25% and 43% respectively, to average income. On the other hand, for small landholders, regular employment (13%) and daily wage labour (8%) were the key subsidiary sources of income, while a marginal income was generated from forest-based

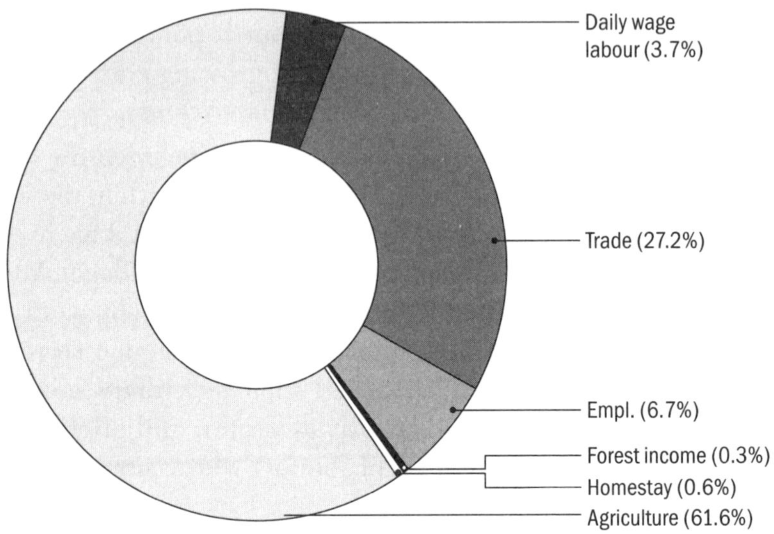

Figure 1 *Income pattern by sources in the study village*

activities such as collection of non-timber forest products (NTFPs) and tourism.

Conservation and Livelihood Linkages

Forest quality and biodiversity values

The historical account traced through the timeline exercise, starting from as early as the mid 1960s, reported many stages of transitions in the forest quality and associated livelihoods of the Hmong inhabitants. Forest quality that was rated '3' on a scale of 1 to 3 (poor to good) until the mid 1970s was reduced to '1' by the late 1980s. This was due to continued destruction of forests, first for opium cultivation on the upper slopes and later, once a ban on opium cultivation was imposed in 1982, for the cultivation of rice, corn, cabbage, and lychee. The negative impact of forest destruction was reaffirmed by the respondents during participatory discussions. Acute shortage of water, high rates of soil erosion, silting of water courses, frequent flash floods during the rainy season, deteriorating quality of drinking water, and reduced availability of forest products were some of the effects of forest destruction listed by the villagers.

Table 1 *Income patterns across different landholders in the BMSM study village*

Type of farmers	*Average landholding (rai)*	*Average annual income (in Thai baht) from different sources*						*Total average income (in Thai baht)*
		Agriculture	*Daily wages*	*Trade*	*Employment*	*Forest*	*Tourism*	
Small	7.9 (73.0)	86 125 (8.1)	9613 (4.9)	5750 (12.9)	15 254 (0.0)	0.0 (1.1)	1250	117 991
Medium	13	134 500 (70.0)	7500 (3.9)	47 500 (24.7)	0.0 (0.0)	3000 (1.6)	433 (0.2)	192 075
Large	20.8	145 340 (49.9)	1800 (0.6)	125 200 (42.9)	18 000 (6.2)	0.0 (0.0)	1200 (0.4)	291 540

Note Values in parenthesis represent the percentage of the total average income

In such a scenario, livelihood was all the more uncertain and difficult, because the villagers had to pay for the products that they used to earlier collect from the forest for free.

Delineation of DSNP only added to the difficulty, as access to the forest resources was further restricted and agricultural expansion to new areas was strictly banned. Moreover, the local inhabitants not only feared eviction after notification of DNSP, but also got into frequent conflicts with the park authorities over the use of resources, including lands for cultivation. The only way out from such a situation for the villagers was to change their image of forestry destroyer and win the confidence of the authorities to secure their claim to remain inside the park. Therefore, they followed the community forestry approach and decided to proactively reforest the degraded areas on the upper watershed forests. Later, technical support from the Forest FORRU-CMU helped to biologically diversify the restoration process (Elliott and Kuaraksa 2008; Goosem and Tucker 1995; FORRU 2006). The combined efforts yielded positive results and the forest quality was progressively rated as '2' by the respondents during the late 1990s and '3' at the time of this study (2009). The qualitative changes reported by the villagers about improvements in the biodiversity have been quantitatively substantiated by a few recent research studies (Elliott and Kuaraksa 2008; Sinhaseni 2008). The combined efforts of the villagers, the FORRU, and DSNP resulted in the regeneration of diversified planting stock and increased the diversity of animal and bird species. The planting of 29 framework tree species resulted in more than 70 'recruit' (non-planted) tree species re-colonizing the restoration plot within eight years (Sinhaseni 2008).

Some of the wild animal species reported to have been spotted frequently in the restored forest were wild boars, barking deers, civets, jungle fowls, and pheasants. The bird species richness reported impressive improvement, from 30 before planting to more than 80 within six years (Toktang 2005). Noticeable reduction in the hunting of wild animals and birds has helped in improving their population. This was reflected from the results of a perception study, according to which 47% of 17 respondents reported a moderate decline, while the remaining 53% reported a high decline in hunting incidences.

Environmental services

Qualitative information was collected from the respondents on important ecosystem functions of the regenerated forest such as soil erosion control, water availability in the streams, soil moisture retention in the downstream agricultural lands, forest fire control, biodiversity richness, and availability of NTFPs. In general, villagers perceived a moderate-to-high improvement in most of the ecosystem function values (Figure 2). Respondents provided positive feedback on the impact of improved environmental services, particularly on agricultural productivity. On a scale of 1 to 3 (low to high), 27% respondents rated low, 60% respondents rated moderate, and 13% respondents rated high improvement in agricultural productivity. Although quantification of the improved productivity was not possible due to the short duration of the study, some of the respondents indicated an improvement in the range of productivity between 10% and 30% increase, particularly in the case of lychee. To some extent, they related this change to the increased availability of water for irrigation due to rejuvenated streams and the improved soil moisture in the downstream cultivated lands.

Forest quality and income patterns

It was interesting to note that despite improved availability of forest resources—and a significant proportion of the respondents (76%)

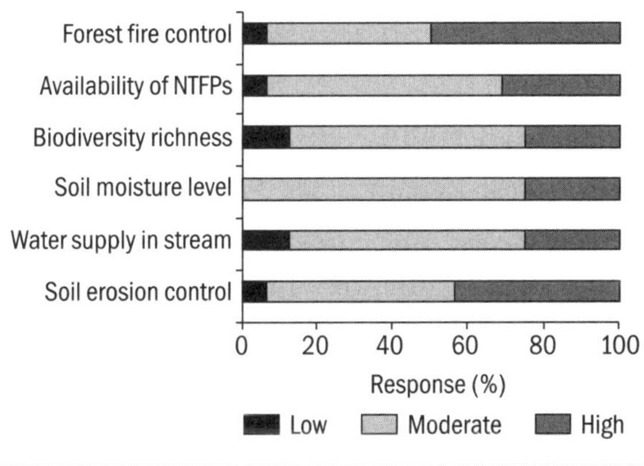

Figure 2 *Perception of the respondents on various parameters of the restored ecosystem*

reporting their dependence on these resources—the direct and indirect income from the forest sector was marginal. Forest produce collected by the villagers included bamboo shoots, wild banana fruits and flowers, mushrooms, wild eggplants, other edible fruits, flowers, and a few medicinal herbs. Out of the 14 forest-dependent house holds, only one respondent reported earning cash from the sale of NTFPs; another five earned income from limited tourism activities by renting out home-stay facilities to nature tourists and students who came to the village for research work or leisure. When asked about the links between the improved forest, employment opportunities, and household income, 37.5% respondents did not perceive any improvement in employment opportunities due to better forest quality, while the same proportion of the respondents reported moderate improvement, and 25% reported low improvement. Plantation programmes, forest fire control, weeding, and fertilizer applications are some of the activities that provide occasional employment to the villagers, but these activities do not seem to translate into significant cash earning. One fourth of the respondents did not report any addition to household income due to such activities, while the rest reported a low (50%) to moderate (25%) contribution. Nevertheless, there was a unanimous perception across all the respondents on the importance of the forest as a livelihood support system, and more importantly, on transforming their image from forest destroyers to forest conservators and thus winning the confidence of the authorities and assuring their stay inside DSNP.

Conservation awareness and regulatory practices

The forest restoration process that started almost two decades earlier also helped to generate awareness among the local people and inculcated a sense of self-regulatory practice with respect to resource harvesting from the forest. A perception study reported 50% respondents with moderate awareness, while the rest 50% were reported with high awareness about forest conservation. A ban on hunting and timber felling, a regular watch and ward against forest fires, participatory plantation programmes, and student research programmes are some of the activities that have contributed significantly to improving forest quality and helped transform the image of the villagers and win the confidence of the park authorities. Currently, there is no formal institution, but a conservation club in the village helps raise awareness and motivates the villagers to participate in plantation programmes, fire brakes, wards, and watches. However, lack of funds or a proper institutional structure was reported as a major obstacle in making further conservation programmes.

Institutional linkages

Focused group discussions with the villagers listed DSNP, FORRU-CMU, and Tambon Administrative Organization (TAO) as key organizations with respect to forest and land management, besides a couple of other projects such as the Upland project and Royal project, which provided local villagers with some regular income. The perceptions study of the villagers on their relationship with the key organizations mentioned above reported a noticeable improvement in this relationship (Figure 3). Respondents linked these positive changes to the constant improvement in the forest quality and, more importantly, to their transformed image. They expressed a sense of pride when visitors, especially school students, researchers, and forest management practitioners, visited BMSM to learn about forest restoration and its livelihood benefits.

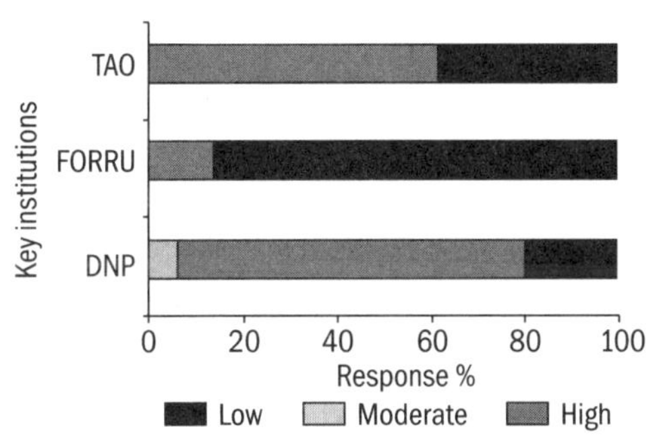

Figure 3 *Perception on the relationship of the local villagers with key institutions*

State of basic infrastructure

The improved relationship with key departments, particularly with TAO, helped in the creation of basic infrastructure such as a school, health facilities, and road connectivity. The villagers acknowledge the contribution of these facilities to their livelihood security. They acknowledged round-the-year road connectivity to the business town, Chiang Mai, as a particularly important development. On the one hand, this increased market access for the villagers to sell their agricultural produce, and on the other hand, it opened up opportunities for the trade of handicrafts and jewellery items for additional income. This trend is reflected in the fact that household income pattern reported trade as the second highest contributor to the average annual income for medium to large landholders. In addition, better connectivity to Chiang Mai also provided opportunities for the younger generation to go for higher education and facilitated more frequent visits to the village by nature lovers, tourists, and students, offering income opportunities to the limited number of households.

DISCUSSION

Community forestry has been advocated by many practitioners as an effective management tool to achieve the twin objectives of forest conservation and livelihood security (Malla 2000; Gilmour, Malla, and Nurse 2004). This is further substantiated by the progressive evolution of Community Forest Management (CFM) from isolated pilot initiatives in a few countries such as India, Nepal, and the Philippines in the early 1970s to a well-established, institutionalized process at present in more than 100 countries worldwide. The case study of Ban Mae Sa Mai further strengthens the concept of community forestry and its role in securing the livelihoods of the local communities. Although regenerated forests may not have contributed significantly in terms of direct cash income, a sizeable proportion of the households highlighted the role of regenerated forests in securing their livelihood. More importantly, the improved relationship with the park authorities also helped them secure their place inside the national park, which was otherwise threatened by the reputation of Hmong as forest destroyers. The important role played by the villagers in conserving the degraded forests and rejuvenating water sources helped improve the environmental services and their social image. Such a transformation also indicates the necessity and stewardship of the local people towards forest conservation for their survival, and, moreover, for ensuring their continued presence inside a protected area.

CONCLUSION

Mainstreaming CFM in national forest programmes in many countries and establishing links between 800 million of the world's poorest people and the forests are indications of the growing importance and acceptance of CFM as an approach to address widespread forest loss and its consequent environmental degradation and negative impact on rural livelihoods. The success of CFM in increasing the area and quality of forests, especially in countries such as Nepal and India, is now reasonably well documented (Kanel 2004; Pokharel and Nurse 2004; Bahuguna 2001). Clear evidence from other countries in the region is still scarce and needs greater research and documentation to provide much needed impetus for policy formulations at the national and regional levels to address conservation and livelihood security more effectively. The emerging challenges of addressing distributional issues such as equity, rights, governance, and poverty alleviation, which have become mainstream objectives of rural development programmes in most countries, further emphasize the need to understand the links between forest conservation and livelihood security.

ACKNOWLEDGEMENTS

The author would like to sincerely thank the Asian Scholarship Foundation for providing the fellowship to conduct this case study. The author would also like to thank the RECOFTC for hosting him as an Asia Fellow. The field support provided by Forest Restoration Research Unit of Chiang Mai University, particularly by Prof. Steve Elliot and his students Sarah and Suhasini, is sincerely acknowledged. Last but not the least, the cooperation and support from the villagers during this study is much acknowledged and appreciated.

REFERENCES

Bahuguna, V. K. 2001. Production, protection and participation in forest management: an Indian perspective in achieving the balance, pp.1–15. In *Forests in a Changing Landscape.* [16th Commonwealth Forestry Conference Proceedings, Freemantle, Australia]

Dearden, P., and W. Tungittiplankorn. 2002. Biodiversity conservation and cash crop development in northern Thailand. *Biodiversity and Conservation* 11:2007–2025

Delang, C. O. 2002. Deforestation in northern Thailand: the result of Hmong farming practices or Thai development strategies. *Society and Natural Resources* 15:483–501

Elliot, S., and C. Kuaraksa. 2008. Producing framework tree species for restoration of forest ecosystem in northern Thailand. *Small Scale Forestry* 7:403–415

FORRU (Forest Restoration Research Unit). 2006. *How to Plant a Forest: The Principles and Practice of Restoring Tropical Forests.* Thailand: Department of Biology, Chiang Mai University

Geddes, W. R. 1976. *Migrants of the Mountains: The Cultural Ecology of the Blue Miao (Hmong Njua) of Thailand.* Oxford: The Clearndon Press

Gilmour, D., Y. Malla, and M. Nurse. 2004. *Linkage between Community Forestry and Poverty.* Bangkok: Regional Community Forestry Training Center for Asia and the Pacific

Goosem, S. P., and N. I. J. Tucker. 1995. *Repairing the Rainforest: Theory and Practice of Rainforest Re-establishment in North Queensland's Wet Tropics.* Cairns, Australia: Wet Tropics Management Authority

Kanel, K. R. 2004. Twenty-five years of community forestry: contribution to millennium development goals. In *Proceeding of the Fourth Workshop on Community Forestry*, edited by K. S. Kanel, P. Mathema, B. R. Kanel, D. R. Niaurla, A. R. Sharma, and M. Gautam, December 2004, Community Forestry Division, DOF, Kathmandu, Nepal

Lieorungruang, V., and N. Muangyai. 2008. Cultural landscape analysis of Doi Suthep-Pui National Park, Chiang Mai. *Silpakorn University International Journal* 8:39–72

Malla, Y. B. 2000. Impact of community forestry on rural livelihoods and food security in Nepal. *Unasylva* 202:37–45

Mottin, J. 1980. *History of the Hmong.* Bangkok, Thailand: Odeon Store

Pokharel, B., and M. Nurse. 2004. Forest and peoples' livelihoods: benefiting the poor from community forestry. *Journal of Forestry and Livelihoods* 4(1):19–29

Renard, R. D. 1997. The making of a problem: narcotics in mainland Southeast Asia. In *Development or Domestication? Indigenous Peoples of Southeast Asia*, edited by D. McCaskill and K. Kampe, pp. 307–328. Thailand: Silkworm Books

Silori, C. S. 2010. Analysis of biodiversity conservation based poverty alleviation initiatives in Thailand. [Final report submitted to the Asian Scholarship Foundation, Bangkok]

Sinhaseni, K. 2008. Natural establishment of tree seedlings in forest restoration trials at Ban Mae Sa Mai, Chiang Mai Province. [MSc thesis submitted to Forest Restoration Research Unit (FORRU), Science Faculty, Graduate School, Chiang Mai University, Thailand]

Thailand Development Research Foundation. 1997. *Management Plan of Doi Sutep-Pui National Park*, vol. 3, B. E. pp. 2541–2550. Bangkok: Thailand Development Research Foundation

CHAPTER 10

Increasing Livelihood Security through Kuth Cultivation in Tons Watershed, Garhwal Himalaya

Arun K Badoni

INTRODUCTION

Organizations around the world have been advocating sustainable development practices in every field of human development. However, the result of this advocacy is insignificant, especially in identifying and establishing the options for sustainable development due to multiple complexities. This is particularly true for a densely populated country like India, where millions of people live in extreme poverty and deprivation. Medicinal and aromatic plants (MAPs) are seen as a potential source to help address some of such concerns, since MAPs have been playing a vital role as a life-supporting system of contemporary civilizations in maintaining good health and well-being of mankind. In India the tradition of using medicinal plants to cure diseases has existed for more than 5000 years. The Himalayan mountains have particularly been a source of rich diversity of medicinal plants and associated knowledge. Therefore, the arguments in favour of Ayurveda, meaning the knowledge of longevity, originating from the Himalayas are not strange.

In particular, the state of Uttarakhand in the Himalayas is well known for its valuable MAP species. 'Charak-ka-Daanda' is located in the Garhwal region of Uttarakhand from where Charak, the great exponent of medicinal science, described panacea for several incurable diseases. Unfortunately, during the last couple of decades, phytodiversity is facing several man-made threats such as destruction or modification of habitat, overexploitation of useful plant species for commercial, scientific, and educational purposes, overgrazing by domesticated animals beyond the carrying capacity, change in arable farming, ploughing of old grasslands, urbanization and other so-called developmental activities, neo-tourism, mining and quarrying, as well as pressure from introduced plants. Apart from this, several natural factors, such as floods, droughts, diseases, critically low population of some species, lack of pollinators, and invasion by exotics or other aggressive species add to the pressure. Besides natural and man-made developmental activities, commercial exploitation of plants has caused depletion of specific plant species (Kala 2000, 2005a, 2010).

Considering the IUCN criteria, about 121 species have been recorded in the Red Data Book of Indian plants from the Himalayan region, of which 17 are medicinal plants (Nayar and Shastry 1987). There are other species whose population is fast decreasing due to their multiple uses and high demand in the pharmaceutical industry. Some of the notable species are *Aconitum* spp., *Allium* spp., *Saussurea* spp., *Dactylorhiza hatagirea*, *Meconopsis aculeata*, *Podophyllum hexandrum*, *Megacarpaea polyandra*, *Swertia chirayata*, *Ephedra gerardiana*, *Didymocarpus pedicillata*, *Mahonia napaulensis*, *Berberis* spp., *Jurinea macrocephala*, and *Begonia josephii*. Hence, it is not enough to maintain databases regarding the status (rare, threatened, vulnerable, endangered, and/or near to extinction) of specific species; it is also necessary to take effective steps to protect the vanishing MAPs along with the preparation of the 'Green Data Book' which contains the list of plant species having reasonably good population size. Therefore, the practice of ex-situ MAPs cultivation is urgently needed.

Until some years ago, the alternative or traditional systems of medicine including Ayurvedic system were seen by many as an unrefined medical system. However, in recent times perceptions seem to be changing now and people have started calling it the time-tested traditional system (Badoni 1990; Badoni and Badoni 2001). With globalization, the importance of this system is being realized around the world creating a greater pressure on the resource base of this system. To save the natural herbs of the Himalayas for the welfare of the future generation of humanity and at the same time meet the requirements of the present generation, ex-situ conservation through cultivation of MAPs is one of the best options. The traditional ISM (Indian System of Medicines) consisting of Ayurveda, Unani, and Siddha have their own prescription for the treatment of diseases. The increasing demand for medicinal plants by the pharmaceutical industries has created a scarcity of raw materials (Kala 2006a, 2009). The absence of national standards and quality control of crude drugs has resulted in adulterated and substandard products. The only solution lies in the scientific cultivation of MAPs. Keeping this view in mind, cultivaton of Kuth (*Saussurea lappa*) is being practised and promoted in the Garhwal region of Uttarakhand. This chapter deals with the various issues related to Kuth cultivation.

UTTARAKHAND: A STATE OF MEDICINAL AND AROMATIC PLANTS

The state of Uttarakhand was carved out from Uttar Pradesh and notified as the 27th state of the Indian Union on 9 November 2000. It shares international boundaries with Tibet in the north and Nepal in the east along the snow-fed Kali River, called Sharda after Tanakpur. In the west, it is bounded by Himachal Pradesh and the Shiwalik plains, while Uttar Pradesh lies on its southern boundary. Uttarakhand comprises the Kumaon and the Garhwal divisions which were carved out from Uttar Pradesh.

This newly created state is a repository of rich vegetational and floral wealth, including medicinal and aromatic plant species. Its varied zones—alpine, sub-alpine, temperate, and subtropical—provide a remarkable range of habitats, which can be seen in the diversity and abundance of species that flourish here. Since time immemorial, the natives have been thriving on these natural resources which were an integral component of this self-sustained system (Kala 2004,

2005b, 2006b). Limited use of forest produce was the only means of earning livelihood for these people. Due to several reasons, many MAPs in Uttarakhand are reported to be on a continuous decline in terms of number and abundance. There is a growing concern about the need to conserve these plant resources due to the growing appreciation of their ecological importance and economic utility. In the recent past, this concern has found expression in the form of several committees getting established including a task force on the Himalayan medicinal plants (Kala 2008). Various programmes have been launched and some measures have been taken to conserve the natural resources, including MAPs. A negative list of plants is prepared to regulate the overexploitation of important species.

However, indiscriminate exploitation of some prohibited species continues unchecked because of the close nexus between the authorities responsible for the execution of the concerned laws and the stakeholders engaged in illegal trade as the latter are quite influential in terms of political links, money, and muscle power. There are concerns of increasing problems of inhabitants of the protected areas on the use of natural resouces. It is evident that they were once the customary owners of all the resources there but are now denied even their basic rights of survival as they are prevented from using forest produce in the name of conservation under the Wildlife Protection Act.

In the absence of any stable agricultural system and employment opportunity in any other sectors, there is mass migration of males to urban areas in search of better livelihood—a serious sociological problem that exists in the area. Women left behind in the rural areas have no alternative but to carry out labour intensive and less productive agricultural activities and work as wage labourers whenever they get a chance to do so. Considering the plight of these women and the sociological problem of the outward migration of males from the area, it is the need of the hour that some alternate effective measure of income generation is created with immediate effect.

The demand for Ayurvedic formulations is increasing at both the domestic and the international levels. The international market for medicinal plant-based products is estimated to be growing every year at 7%. At present, the domestic and international markets' requirements are being fulfilled by legal or illegal exploitation of plants from their wild habitat. Therefore, a number of species are on the verge of becoming endangered and many have reached the point of extinction. After the creation of Uttarakhand, it is felt that the cultivation, conservation, and marketing of plant-based products will play a vital role in the economic sustainability of the state.

COLLECTION PRACTICES OF MEDICINAL AND AROMATIC PLANTS

The current cultivation of medicinal plants in the state is still in the primitive stage due to the lack of suitable agro-technology and poor marketing facilities (Kala and Silori 2010). Most of the areas suitable for natural growth as well as commercial cultivation of plants are deep in the interior, devoid of any transportation system. It is reported that the Western Himalaya alone contains about 50% of the plant drugs mentioned by the British pharmacopoeia and the region caters to about 80%, 46%, and 33% of the Ayurvedic, Unani, and allopathic systems of medicine, respectively. Almost all these systems, at present, depend on collections from the wild, procured legally or illegally, which have endangered the regeneration potential of some taxa and reduced the availability of others to a great extent.

Collection from the Wild

The wild continues to be the major source of supply. Legal collection is organized through the Bhaishaj Sangh and Forest Corporation. The Sangh does not undertake collection directly, but gives the responsibility to local contractors. Many of these contractors also act as middlemen and traders. The people involved in the actual collection are either wage labourers or farmers with small landholdings. Many of the collectors are not from the state and, in many cases, are from Nepal. There is a general feeling that this system of contracting contributes to both illegal and unscientific collection. There have been recent attempts to reduce the role of contractors in the collection process. For example, the contract to collect is now given only to village community organizations called *samooh*. However, even this system does not seem to be very effective since the contractors have also set up their own *samoohs*, thus continuing to maintain their control over collection activities.

An important reason why contractors and middlemen exercise such strong control is the collectors' dependence on loans. This practice is widespread. As most collectors are poor, they need to borrow money from the contractors. This keeps the collectors tied to the local contractors. Breaking the hold of the contractors over the collection continues to be an important challenge.

Illegal Collection

Illegal collection continues to be common. Middlemen and local contractors are often involved in both legal and illegal collections. This allows them to market illegally collected material along with the legal material.

Large-Scale Collection

Large-scale collection of plants has led to a significant depletion of important species in some area. Farmers in the Yamuna valley report that, because of depletion, the volume of plant species a person can collect has declined significantly in recent years. According to one collector, earlier he used to collect about 200 g of dry Atish in one day. Now he does not get more than 20–50 g a day.

CULTIVATION OF MEDICINAL AND AROMATIC PLANTS

The Uttarakhand government has a policy to promote cultivation of medicinal plants, motivated by the need to conserve forest resources and increase the income of farmers through agricultural diversification. Some of its components include a relaxation of restrictions on cultivation, development and diffusion of cultivation technologies, supply of planting material, credit, and marketing support. An important premise of the policy is that increased cultivation will reduce the demand for the material collected from the wild. The study, however, suggests that there is little direct connection between collection and cultivation. Most collectors are wage labourers who have no other source of income. Even if cultivation is adopted by a large number of farmers, collection will continue to be carried out by wage labourers. However, if the user industry begins to show a preference for cultivated material, the demand for collected material could decline in time. The following factors, however, prevent farmers from cultivating medicinal plants on a large scale.

Small landholdings: A majority of farmers in Uttarakhand have small landholdings so there is a limit on the area of land that can be converted for medicinal plant cultivation.

Long gestation period: Many medicinal plants can be harvested only after three years or later. This is particularly true for plants grown in high altitude areas.

Lack of farmer-driven agro technologies: Many medicinal plants are difficult to cultivate. This is particularly true for plants grown at high altitudes. The farmers face serious difficulties, as they have no experience in cultivating medicinal plants. There is a clear need for the development of technologies related to cultivation, harvesting, storage, transportation, and quality control. Although there are a number of institutions in Uttarakhand conducting research on medicinal plants, unfortunately, they have limited contribution to standardize agrotechniques of MAPs. This may be due to a number of reasons, which include the following.

Sub-optimal use of limited resources: The resources available for research on medicinal plants are small. In order to use these resources efficiently, it is important to build up a close collaboration among various research institutes. The links between institutes and farmers are weak. Most of the research is being done in laboratories. There is very little on-farm research. This has a number of implications which are as follows.

- Problems faced by farmers are not fully reflected in the research efforts.
- Researchers are unable to take advantage of farmers' experience.
- As farmers are not involved in the research process, diffusion of technology is hampered.

Research institutes do not have resources or experience to undertake a large-scale diffusion of technologies. This gap needs to be filled. Institutes also lack resources to train farmers. Although some efforts have been made to train farmers, the number of such trained farmers is small. The links between research institutes and pharmaceutical industry are also weak. Barring a few exceptions, there is no collaboration between research institutes and industry. There is a general lack of appreciation of the advantages of collaboration in both. The industry is reluctant to support research, as it is not sure of the commercial returns. The pressure of peer group and academic compulsions, on the other hand, still largely dictate the research priorities of institutes.

There is a serious shortage of planting material to set up cultivation in large areas. Many farmers who are interested in cultivation reported difficulties in obtaining suitable planting material.

Supply chain and market-related problems: It is found that there are a large number of stakeholders in the supply chain and include primary collectors and producers, local contractors, middlemen, regional wholesale markets, markets in large centres such as Delhi, and specialized suppliers. In almost all cases, the primary collectors and producers sell to local contractors and middlemen. As they are tied to these agents through debts and other obligations, their bargaining power is negligible. Also, as they have only small amounts to sell, they do not have the option of selling directly to wholesalers.

The medicinal plants sector in Uttarakhand faces a number of market-related problems. Reliable information on demand and prices of MAPs is not available. Considering that government agencies and research institutes have little contact with the industry, this is not surprising. It was found that, in many instances, the prevailing market prices were lower than the information provided by official agencies. Furthermore, the official estimates do not take into account large fluctuations in demand.

Prices are too low to make cultivation attractive. Many farmers feel that the prices they receive do not compensate for the difficulties and uncertainties involved in the cultivation of medicinal plants. The prices are low for two important reasons. Firstly, the low cost of collection (both legal and illegal) puts a downward pressure on the price, making cultivation unattractive. Although accurate information on the cost of cultivation is not available, there is a general feeling that it is higher than the cost of collection. Cultivation requires land, labour, planting material. Also, the returns from cultivation, in many cases, are available only after 3–4 years. On the other hand, the cost of collection largely constitutes labour and the returns are available within a few months. Secondly, as the cultivation of medicinal plants is a comparatively new activity, the yields are low. Therefore, as long as large-scale collection continues and improved cultivation technologies are not transferred to farmers, the latter may be reluctant to undertake the cultivation of medicinal plants.

Since the supply chain is long, with as many as six or seven players involved, the prices farmers receive are considerably lower than those paid by the user industry. However, as the traders play a crucial role in the smooth functioning of the chain, it is unlikely that direct selling to the industry could become a practical option.

Large exporters may find buy-back arrangements attractive, as the requirement of traceability will become increasingly important. At present, it is impossible to trace the origin of any material, as there is little transparency or documentation. Contract farming and buy-back arrangements provide the only practical alternatives to exporters whose customers require traceability. However, as it is still possible to export without traceability, the full impact of this trend is yet to be felt in India.

It is important to stress that medicinal plants originating from Uttarakhand compete with materials from other states in India, and farmers must be approached only when all components necessary for cultivation and marketing are in place. In the past, efforts to popularize cultivation had not geared up to the desired level. For example, a decade ago Bhaishaj Sangh distributed seeds of *S. lappa* (Kuth) to the farmers in Chamoli district without adequate resources to provide technical and marketing support. The result was that the farmers faced numerous difficulties in both, growing and marketing the production. This has caused a feeling of skepticism among the farmers. It will take a lot of effort to get these farmers interested in the cultivation of medicinal plants again.

ROLE OF SHER IN FARMING OF KUTH

The Society for Himalayan Environmental Research (SHER) was founded with the broad objective of bringing about sustainable development of the Garhwal Himalayan region through biodiversity conservation and natural resource management. Directed towards achieving the objective, the society is engaged in carrying out Applied Research (AR) and Participatory Action Research (PAR) on different aspects of the Himalayan environment in general and in-situ and ex-situ conservation of medicinal and aromatic plant species in particular.

The SHER has successfully conducted a state-of-the-art study on species such as *Aconitum atrox*, *Aconitum heterophyllum*, and *S. lappa* to propagate the genetic resources of these plants in controlled conditions, generate appropriate agro technologies for their cultivation in the field, and work out techno-economic viability for their commercial cultivation. Though a long list of evolved scientific practices are available, their application for large-scale cultivation, especially of medicinal plants, is still unexplored because of a number of inherent constraints involved. Before proceeding seriously in this direction, these constraints should be looked at. The most significant constraint is the lack of an adequate amount of germ plasms of the said taxa required for their large-scale cultivation.

Cultivation of Kuth: A Success Story

Kuth is one of the important medicinal plants of the Himalayan region. It is used to cure chronic and foul ulcers, leprosy, leucoderma, pruritis, amenorrhoea, and other health problems. The roots are aromatic, aphrodisiacal, digestive, stimulant, and tonic. Due to several reasons, the population of Kuth has declined severely, resulting in the loss of raw material for preparing medicine, and the valuable knowledge associated with it. Besides, to meet the gaps in the demand and supply of this important medicinal plant, SHER started large-scale cultivation of Kuth in the upper catchments of the Tons watershed in Garhwal Himalaya.The commercial cultivation of Kuth is being operationalized in 20 ha of land within several villages of Panch-gai, Adoer, and Badasu Pattis, including Liwari, Kasla,

Rala, Regcha, Phitari, Satuti, Sawani, Sirga, Dhara, Jakhol, Pawn Talla, Osla, Purti, Sirga, and Saur within the upper catchments of the Tons watershed of Uttarakhand.

In the first phase, more than 750 000 individuals have been raised for the fix plantation of nearly 600 000 plants of Kuth to achieve the goal of large-scale cultivation on the 20 ha land within several villages of the project area. The desired technical know-how pertaining to Good Agricultural Practices (GAP) is being provided to the progressive farmers of the targeted area.

On the other hand, the post-harvesting techniques, appropriate value addition methods, API grading, and proper packaging, labelling, and storage mechanism are being disseminated to the Kuth growers in the study area. It is worth mentioning that the present venture into Kuth cultivation is certainly playing a vital role in increasing the livelihood security of the local populace as well as in making Uttarakhand a real 'Herbal State'. This is evident from Table 1, which tabulates the comparative cost-benefit analysis of existing cash crops vis-à-vis Kuth cultivation.

In order to meet the goal of sustainable development, SHER has been working since its inception to mitigate the problems looming large in the hill society, especially between jobless economic growth and the livelihood security of the poor, between resource extraction and ecological sustainability, and between short-term opportunity and long-term stability. It is felt that development policy, which can help to resolve these major problems, may be sustainable in the long run.

Details of the Activities Conducted

Establishment of nursery

A nursery was established on 3 acres within the upper catchments of the Tons watershed of Mori Block in the Uttarkashi district of Uttarakhand. Several activities were involved in the preparation of the nursery such as ploughing and exposing the soil to the sun in summer to kill any harmful organisms within. In the high altitude areas of the Himalaya, summer ploughing improves soil structure due to alternate drying and cooling; soil permeability is increased

Table 1 *Economic analysis for 1 acre land (20 nalis)*

Species	*Time*	*Input (years)*	*Yield (kg)*	*Output (₹)*	*Net profit (₹)*	*Net profit per year (₹)*
Kuth	2.5	30000 (12000 plants)	1800 kg (150 g per plant)	180000 @₹100/kg	150000	60000
Potato (village away from road)	1	25000	25000	40000 @₹8/kg	15000	15000
Potato (road head village)	1	20000	5000 kg	40000 @₹8/kg	20000	20000
Rajma	1	5000	300 kg	15000 @₹50/kg	10000	10000

by breaking the compacted layers. Tillage improves soil aeration during this season to help the multiplication of microorganisms and decomposition of organic matter, resulting in higher nutrient availability. The increased aeration also helps in degrading harmful allelopathic chemicals exuded by the roots of previous crops or weeds and also helps in reducing soil-dwelling insect pests. In the initial stage, the lignose elements were uprooted and the activities pertaining to the breaking and removal of big boulders were carried out, followed by the preparation of the retaining walls and levelling of the fields. For providing a conducive environment to rapid seed germination and growth, raised seed beds were prepared in which the surface soil was soft, fertile, and free of clods of trashy crop residue; the subsoil was permeable to air and water coupled with adequate drainage and aeration.

The nursery area was repeatedly ploughed and harrowed to reduce the size of soil clods and produce a smooth soil surface and to level the soil. Loosening the soil more than 30 cm below the surface helps plant roots thrive. Double digging was carried out since it adds air deep into the soil and helps roots and microbes grow to create a good soil structure. Double digging is also an effective method for improving the soil in the field with just a spade and a lot of hard work.

Forest litter, FYM, and vermicompost (3:2:1) were used to improve the physio-chemical properties of the soil to get easy emergence of seedlings from crusting soils. Great care was taken for the provision of optimum light, temperature, and aeration of the seed bed to get good stocky plants ready at the required season. To avoid over heating and for making plants soft, seeds were sown in open beds instead of a polyhouse. After sowing, proper mulching was carried out to protect the seed bed from drying, crusting, and cold, as well as to check the run off and discourage weed growth. Green nets were used over the beds at a height of 1 m as shade from the strong sun for the tender seedlings.

For the preparation of nursery beds, a site of more than 0.4 ha of the 3 ha land was selected to get at least eight hours of sunlight per day. The width of the beds was designed keeping in mind the accessibility for operational activities, particulary for the beds located in the middle of the rows. The height of the bed was kept at around 50–60 cm from the ground while the width ranged from 0.75 cm to 1.0 m and the length was as per the availability of land. These were filled with a mixture of high quality top soil compost and manure. On this site, 70 kg of Kuth seeds were sown in June.

Germination observed under controlled laboratory conditions is rarely achieved in nursery beds. A considerable decrease was recorded in the germination of seed due to various field conditions such as soil texture, temperature, and moisture, which were not equally conducive for seed germination. Besides seed germination, the growth and survival of seedlings is of great concern if the species has to be taken under cultivation practice.

The Kuth seeds, which were procured from the mother plants raised at the SHER field station at Shirolatok, were observed for germination potential under various concentrations of GA3 (Gibbrelline). Five thousand seeds per concentration in each species were dipped into 25 ml solution of GA3 under different levels: 20 ppm, 50 ppm, 100 ppm, and 200 ppm. Natural water was used as the germination medium for control. After 48 hours, seeds belonging to each treatment were sown in seed trays and observed for germination (cotyledon stage) and growth of seedlings. A positive correlation between germination percentage and root weight has been observed with the increasing concentration of GA3 treatment up to 100 ppm. In the case of Kuth, GA3 treatments at

100 ppm concentration enhance the germination percentage up to 67% compared to 42% under control (Figure 1).

On the whole, more than 450 000 Kuth seedlings were raised. The seedlings raised in the seed beds were transplanted before being distributed to progressive farmers to set in their cultivated fields. This pricking resulted in more stocky plants with better roots and root branching. More than 300 000 well-developed seedlings were distributed to 50 farmers (6000 each) for transplantation. Because of low light, high humidity, high temperature, and a large quantity of organic fertilizers used, the seedlings often become very succulent and tender. They do not stand transplantation well unless they are hardened before being shifted to the open fields. Withholding the supply of moisture 10 days before the seedlings were taken out from the nursery, the hardening was done.

Awareness camps: dissemination of technologies to the farmers

In order to work out the multiplication as well as large-scale cultivation strategies for Kuth within the project area, awareness camps were organized in the interior villages located in the upper catchments of the Tons watershed. These camps were organized to involve the local people in the mission of sustainable utilization vis-à-vis conservation of threatened medicinal plant species by making

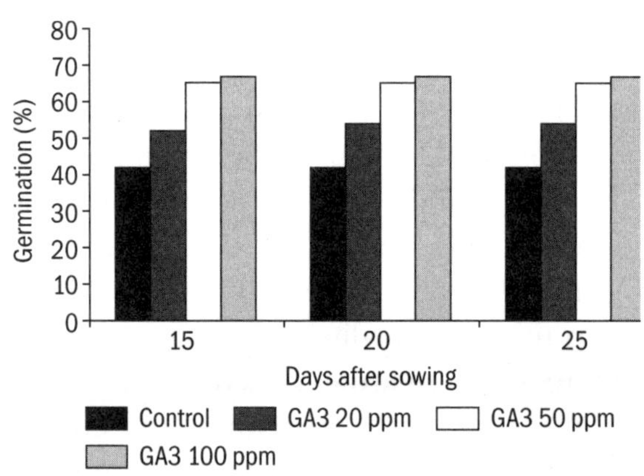

Figure 1 *Effects of hormone GA3 on the seed germination of Saussurea lappa*

Figure 2 *Harvesting of Kuth in Uttarkashi, Uttarakhand*

them aware of the paramount importance of Himalayan medicinal plants, the need of MAP diversity protection, irrational exploitation causing endangerment of this precious herbal wealth, and the associated economic loss. The camps aimed to motivate people for ex-situ conservation (cultivation) of important medicinal plant species for the sustainable development of the area with an emphasis on large-scale cultivation of Kuth for their livelihood security (Figures 2 and 3).

Figure 3 *Tribal community processing Kuth in Uttarkashi, Uttarakhand*

A core group of resource persons including SHER experts, staff members of the project, and forest personnel was constituted to organize the Kuth cultivation camps. Three awareness camps were organized in Kasla, Bhitri, and Datmeer villages. All these campaigns and activities have encouraged farmers in the Tons watershed to cultivate Kuth to increase their livelihood security.

REFERENCES

Badoni, A. K. 1990. An ethnobotanical study of Pinswari community: a preliminary survey. *Bulletin of Botanical Survey of India* 32:103–115

Badoni, A. K., and K. Badoni. 2001. Ethnobotanical heritage. In *Garhwal Himalaya: Nature, Culture and Society*, edited by O. P. Kandari and O. P. Gusain, pp. 127–147. Srinagar, Garhwal: TransMedia

Kala, C. P. 2000. Status and conservation of rare and endangered medicinal plants in the Indian trans-Himalaya. *Biological Conservation* 93(3):371–379

Kala, C. P. 2004. Revitalizing traditional herbal therapy by exploring medicinal plants: a case study of Uttaranchal state in India. In *Indigenous Knowledges: Transforming the Academy*, edited by A. Maretzki, B. L. Farmer, and L. Semali. Pennsylvania: Pennsylvania State University

Kala, C. P. 2005a. Indigenous uses, population density, and conservation of threatened medicinal plants in protected areas of the Indian Himalayas. *Conservation Biology* 19(2):368–378

Kala, C. P. 2005b. Current status of medicinal plants used by traditional *vaidyas* in Uttaranchal state of India. *Ethnobotany Research and Applications* 3:267–278

Kala, C. P. 2006a. Problems and prospects in the conservation and development of the Himalayan medicinal plants sector. *International Journal of Sustainable Development* 9(4):370–389

Kala, C. P. 2006b. Medicinal plants: potential for economic development in the state of Uttaranchal, India. *International Journal of Sustainable Development & World Ecology* 13(6):492–498

Kala, C. P. 2008. High altitude medicinal plants: a promising resource for developing herbal sector. *Hima-Paryavaran* 20(2):7–9

Kala, C. P. 2009. Medicinal plants conservation and enterprise development. *Medicinal Plants* 1(2):79–95

Kala, C. P. 2010. *Medicinal Plants of Uttarakhand: Diversity, Livelihood and Conservation*. Delhi: Biotech Books

Kala, C. P., and C. S. Silori. 2011. Medicinal plants in Nanda Devi Biosphere Reserve: traditional heritage for livelihood and health security of ethnic communities. In *Community-based Biodiversity Conservation in the Himalayas*, edited by A. K. Negi and Y. Gokhale, pp. 75–85. New Delhi: The Energy and Resources Institute

Nayar, M. P., and A. R. K. Sastry. 1987. *Red Data Book of Indian Plants*, vol. I–III. Kolkata: Botanical Survey of India

CHAPTER 11

Case Study of Traditional Agroecosystems in Patharkot Village, Uttarakhand

Narayan Singh and R C Sundriyal

INTRODUCTION

Farming has been a traditional means of livelihood in the mountainous region of Uttarakhand, just as in many other similar regions of the world. Social practices, rituals, festivals, and cultural relations are intertwined very closely with farming operations. However, with the invasion of market forces, the agriculture landscape is changing fast. Like other Himalayan states, the Uttarakhand region of the Indian Himalayas has been rapidly undergoing such changes due to various reasons. Commercialization of agriculture, focusing more on maximizing benefits, has replaced many traditional crop varieties to the extent that some of them are not found in the region any more. Animal husbandry, once an integral and valued part of agriculture, is relegated to secondary importance, as chemical fertilizers have replaced dung manure, and livestock are seen only as meat—in the case of sheep and goats.

An integrated land-use system, comprising forest, agriculture, horticulture, agroforestry, and animal husbandry, plays a vital role in maintaining the economy of the region and providing much-needed subsistence support to the local population (Fonzen and Oberholzer 1984; Misra and Ramakrishnan 1982; Nair 1985; Ralhen, Negi, and Singh 1991; Shah 1982; Sharma and Ambasht 1988; Singh, Rai, Patiram, *et al.* 1989; Toky and Ramakrishnan 1981). This subsistence system has evolved over the years based on trial and error method devised by farmers to meet the demand for food, fodder, fuelwood, and timber. Traditional agriculture practices have been part of such subsistence-based agroecosystems. The Indian Himalaya in general and the Central Himalaya in particular have been a rich source of traditional agriculture crop diversity. In the traditional agroecosystems of Central Himalaya, more than 40 species of food grain are grown, which have been managed by the local farming communities since time immemorial (Maikhuri, Semwal, Rao, *et al.* 1996). However, over the years, there has been a rapid erosion of many such crop varieties and the cultivation and management practices associated with them. In the past, few attempts were made to document such changes. In this study, we have attempted to (1) prepare an inventory of the crops, fodder, and fruiting trees in the study site, (2) analyse the range of inputs in the production of crops, (3) determine the fuelwood consumption pattern, and (4) find out different income groups to understand the interplay of crop diversity and the livelihood support system in the study village.

STUDY AREA

The study was conducted in Patharkot Village situated along River Kosi in the Almora district of Uttarakhand. The total population of this village is 733, out of which 433 are male and 300 are female. The literacy rate of the village is approximately 95%. The total land area of the village is 163.62 ha, out of which Van Panchayat occupies 33.40 ha, 65.97 ha is for agricultural use and 64.24 ha is for land under other uses, including wastelands and settlements. After initial discussion with the local community, we decided to use the appreciative participatory enquiry and planning method for assessing the socio-economic situation of the local communities. The farming system prevalent in the hills is subsistence based. Agriculture is heavily dependent on energy flows from uncultivated lands such as forests and grasslands recycled into manure through livestock (Ashish 1979).

METHODOLOGY

A complete household survey was carried out for documenting cropping patterns, including horticulture crops, and a complete inventory of inputs and outputs in the upland farming systems was generated following Nair (1985). The inputs included (1) labour in terms of bullock days and man days, (2) farmyard in terms of quantities and chemicals (if used), (3) seeds, (4) yield of edible products and by-products. An estimation of the actual amount of fuelwood and fodder consumed was carried out on the basis of survey and regular observations in the villages. The source and supply of fodder for animals was quantified as a contribution of agricultural by-products, forest and tree fodder from agroforestry by estimation in the villages, and by measuring the daily ration of food concentrates given to the animals by each household (Sundriyal, Rai, Sharma, *et al.* 1994). The firewood used by the villagers was quantified by recording the amount of wood used from a pre-weighed amount given to sampled households across different socio-economic categories. We analysed the data for different socio-economic groups in the village to know the income patterns, average agronomic yield, fodder, firewood, fruit production, and the input-output ratio in agriculture.

STRUCTURE OF THE SYSTEM

Cropping System

The cropping system of Patharkot is rainfed and traditional. Farmers grow mainly cereals and pulses. Since the average landholding per farmer is very low, productivity is not enough to sustain their livelihoods throughout the year. Cultivation of cash crops and horticulture crops was not reported. Out of 53 families, 51 are marginal landholders and 2 are small landholders. The average landholding of a farmer in the village is 0.45 ha. A two-year cropping cycle is prevalent in the region. Farmers grow mainly two crops during this period. The land is divided into approximately two parts called *Sari*, and each part is cultivated alternatively. Wheat (*Triticum aestivum)*, paddy *(Oryza sativa)*, mustard *(Brassica compestris*), mandwa *(Eleusine coracana)*, and mixed pulses are the main crops. The average agronomic yield of a farmer in the village is 88 kg (Figure 1).

Livestock

Animal husbandry is an integral part of agriculture cultivation. The animals not only provide milk and other products, but also serve as drought power in the fields. The farmer therefore grows crops that also meet the fodder requirements of the animals, while animal dung is used to fertilize the soil that helps in growing crops. However, over the years, the crop mix has changed noticeably and agricultural production seems to be declining. Such a situation is causing difficulties for the farmers in maintaining large number of animals. This, in turn, is affecting the productivity due to reduced supply of organic manure to the fields. Crop residues supply fodder but due to changes in cropping patterns recently, farmers find it very difficult to sustain the animals and improve their productivity. The most serious problem is the lack of green forage, particularly during winter, which causes deficiencies of protein and vitamins and results in low milk production, high mortality, shortened breeding span, and decreased working capacity in bullocks.

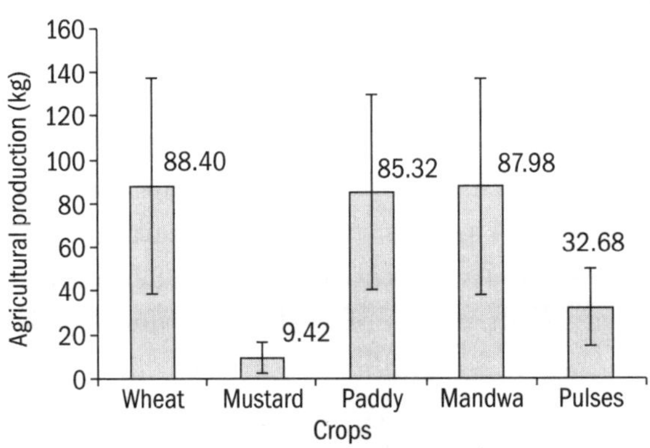

Figure 1 *Productivity of different crops*

The animal species and races in Patharkot are very traditional and low in productivity. The average livestock holding is dominated by goats and cows, while a few families also own buffaloes and bullocks (Figure 2). It is necessary to undertake land rehabilitation programs on community and privately owned degraded land by planting multi-utility trees with people's participation. Also, other management interventions such as controlled grazing and reseeding of legumes to improve the carrying capacity of the range lands are needed. The use of trace risers can be made for cultivation of perennial grasses and fodder trees (Palni, Maikhuri, and Rao 1998).

Horticulture

Within the context of sustainable mountain agriculture, it is now well acknowledged that horticulture plays a vital role in terms of resource base availability, resource use/management practices, and production flow. The mountain environment and resource base offer physically more conducive circumstances for horticulture development (Verma, Awasti, and Agnihotri 2000). Despite the prevalence of horticulture, its development in the region has not been very promising. Something similar was reflected from the study village. On the basis of primary survey and observation of the study village, we reported eight types of fruit trees, dominated by citrus species. The annual productivity of all the fruit crops was 680 kg (Figure 3).

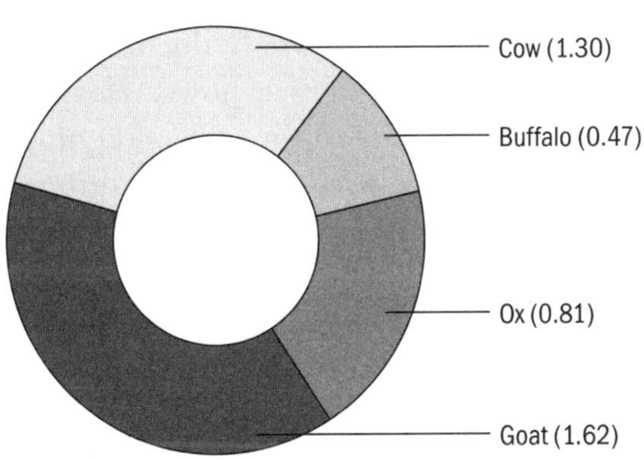

Figure 2 *Average livestock holding per family in Patharkot*

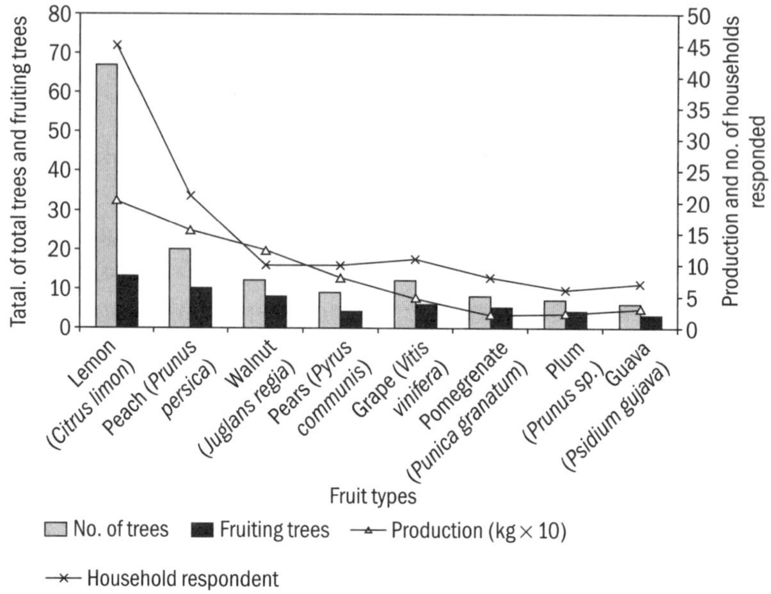

Figure 3 *Horticulture productions in Patharkot Village*

Fuelwood

Among the natural resources of Uttarakhand, forests are the most important, both economically and environmentally (Sati 2005). The forest is the main source of fuel in the state. The use of fuelwood is the predominant form of energy for cooking. Firewood is collected by men, women, and children from nearby forested areas. At higher elevations, people collect firewood only during winter and store it in heaps for the whole year, whereas at lower elevations, collection is done throughout the year. Due to the collection of a huge amount of firewood, forests near the villages are subjected to rapid degradation and overexploitation. A very small fraction of the firewood comes from agriculture fields (Sundriyal, Rai, Sharma, *et al.* 1994). The total forest area for fuel collection for the village is 33.40 ha, which averages 0.63 ha per household. According to the primary survey of the village, fuel consumption is 418.86 MT, whereas the annual fuel availability is 211.03 MT, thus having a deficit of 207.83 MT. Fuel consumption per family is 52.53 kg (summer: 17.28 kg, winter:

35.25 kg). The major fuelwood species are *Quercus leucotrichophora, Pinus roxburghii, Myrica esculanta,* and *Pyrus pashia* (Table 1).

Fodder

In the Himalayan region, domestic animals provide the main drought power for the agriculture system. They also process crop residues, provide essential organic manure, and generate farm income when they are sold (Thapa, Joshi, Sherpa, *et al.* 1991). The domestic animals depend on the forest as well as on agricultural residues. The main type of fodder, sources of fodder area under fodder production, and total production arc quantified by regular observation and survey in the village. From May to July, the maximum fodder comes from agricultural and forest areas and is 197.58 MT (on the basis of dry matter). The villagers collect dry grasses from grass fields locally known as *mange.* Fodder collected from these grasslands is 62.70 MT, while the other sources of fodder are agricultural residues and trcc spccics. The main tree species used for fodder are *Grewia optiva, Celtis australis,* and *Q. leucotrichophora.* The total available fodder is 281.76 MT but the total consumption is 402.72 MT, reporting a deficit of 207.83 MT (Table 2).

Income Groups

As mentioned earlier, the agriculture pattern of the study village is totally traditional and there are no practices of cultivating cash crop in the villages. On the basis of the primary survey, we found that there were five income groups in the village, including 27 families whose major occupation was agriculture. Another 14 families of ex-army persons were having regular income through pension and

Table 1 *Fuelwood supply in Patharkot Village*

Van Panchyat area	33.40 ha
Total household	53 (Family)
Forest area per household	0.63 ha
Fuel requirement per year	418.86 MT
Availability per year	211.003 MT
Deficit	207.83 MT
Major fuelwood species	Banj Oak, Chir Pine, Kafal, Mehal

Table 2 *Fodder demand and supply status in Patharkot Village*

Fodder	*Sources*	*Area (ha)*	*Quantity (dry matter) (tonnes)*
Green grasses	Agricultural fields and forest	15	197.58
Dry grasses	Grasslands (*mange*)	9.04	62.70
Tree fodder	Agricultural fields	NA	3.91
Agricultural residues	Agricultural crops	15	6.99
Oak leaves	Forests (*Bajani*)	33.04	10.78

agriculture was a subsidiary income source for them. Nine families had regular jobs in the government and private sector with agriculture as subsidiary occupation and remaining two families were earning their income from daily wage labour with no agriculture land. Household income was reported maximum for the families having regular jobs in the government and private sector, as shown in Figure 4.

Input and Output Ratio in Agriculture Crops

The two-year cropping cycle is prevalent in the study village. Wheat, *ragi*, mustard, paddy, and pulses are the main crops of the village. The land is divided approximately into two parts, as mentioned before. One part remains fallow while the other is cultivated. The average landholding of a farmer in the village is 0.45 ha and the average annual inputs are as follows. The farmer uses 70 days of labour, 10 days of animal labour, 3395 kg of farmyard manure, and 43 kg seeds of different crops. As an output, he gets 88 kg wheat, 10 kg mustard, 85 kg paddy, and 33 kg pulses. In terms of monetary value, the cost of his total inputs works out nearly ₹16 000, while the value of his output is around ₹5000, a net loss of ₹11 000 (Figure 5).

CONCLUSION

Traditional agriculture is prevalent in the Central Himalayan region. Farmers mainly grow cereal crops. Wheat, *ragi*, paddy, mustard, and pulses are the main crops of this region. Nowadays, traditional agricultural systems are unable to sustain villagers, which is why the practice is on the decline. Farmers get low outputs compared to

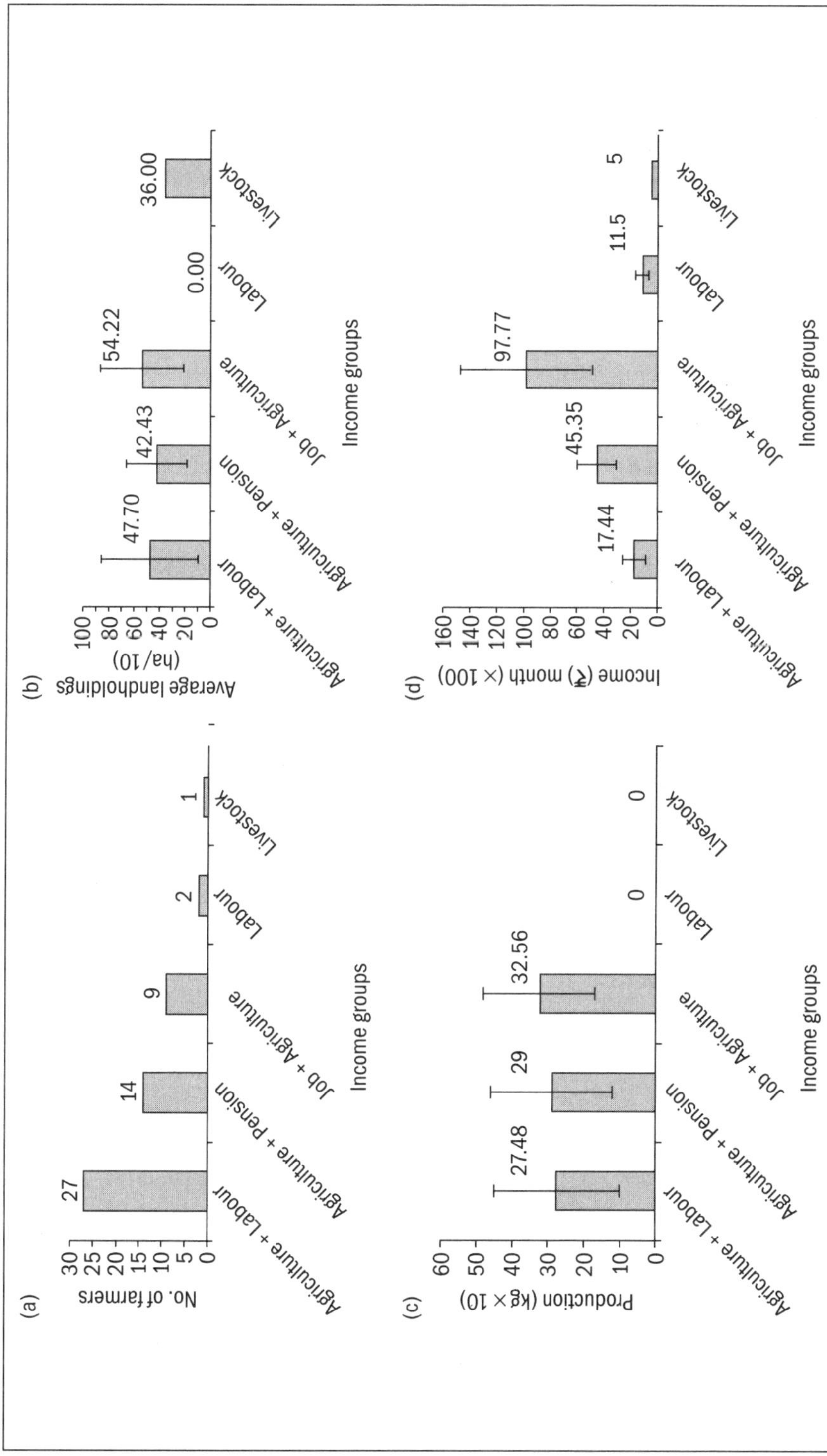

Figure 4 *In Patharkot village (a) income groups, (b) average landholdings, and (c) average agronomic yield, and (d) average income per month*

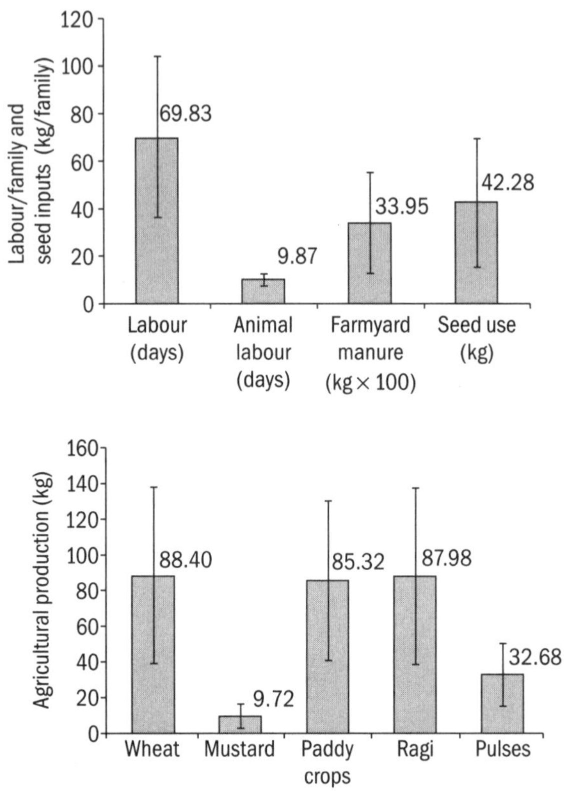

Figure 5 *Input and output ratio in agriculture*

inputs, so agriculture is becoming a less attractive occupation and forcing them to migrate to the lower plains.

A huge deficit in fuelwood and fodder is another major issue of livelihood security among the local people. It is important to emphasize that proper interaction between the villagers, the agriculture department, NGOs, institutions, and government agencies be promoted. A proper and people-centric approach should be implemented in the village according to the needs of the farmers in the different sectors. The formulation and implementation of a regional policy should further improve productivity and conservation in these agroecosystems.

REFERENCES

Ashish, M. 1979. Agriculture economy of Kumaon hills: threat of an ecological disaster. *Economic and Political Weekly* 14(25):1058–1060

Fonzen, P. F., and E. Oberholzer. 1984. Use of multipurpose trees in hill farming system in western Nepal. *Agroforestry Systems* 2:187–197

Maikhuri, R. K., R. L. Semwal, K. S. Rao, and K. G. Saxena. 1996. Traditional crop diversity for sustainable development of central Himalayan agroecosystems. *Indian Journal of Sustainable Development and World Ecology* 1:56–68

Misra, B. K., and P. S. Ramakrishnan. 1982. Cultivation and use of lesser-known plants of food values by tribals in north-eastern India. *Agriculture Systems* 9:57–72

Nair, P. K. R. 1985. Classification of agroforestry systems. *Agroforestry Systems* 3:97–128

Palni, L. M. S., R. K. Maikhuri, and K. S. Rao. 1998. *Conservation of the Himalayan Agroecosystems: issues and priorities*, pp. 263–291. [Report on the International Meeting on Himalayan Eco-regional Co-operation]

Ralhen, P. K., G. C. S. Negi, and S. P. Singh. 1991. Structure and function of the agroforestry system in Pithoragarh district of Central Himalaya: an ecological viewpoint. *Agricultural Ecosystem Environment* 35:283–296

Sati, V. 2005. Natural resource condition and economic development in Uttaranchal Himalaya. *Indian Journal of Mountain Science* 2(4):336–350

Shah, S. L. 1982. Ecological degradation and future of agriculture in the Himalayas. *Indian Journal of Agricultural Economics* 37(1):1–22

Sharma, E., and R. S. Ambasht. 1988. Nitrogen accretion and its energetics in the Himalayan alder. *Functional Ecosystem* 2:229–235

Singh, K. A., R. N. Rai, Patiram, and D. T., Bhutia. 1989. Large cardamom (*Amomum subulatum* Roxb.) plantation: an age old agroforestry system in Eastern Himalaya. *Agroforestry System* 9:241–257

Sundriyal, R. C., S. C. Rai, E. Sharma, and Y. K. Rai. 1994. Hill agroforestry systems in south Sikkim. *Agroforestry Systems* 26:215–235

Thapa, B., L. Joshi, S. L. Sherpa, I. B. Karki, R. K. Kulsi, Y. N. Jha, and M. P. Mainali. 1991. Tree planting and livestock don't mix... or do they? *Agroforestry Today* 3(2):11–13

Toky, O. P., and P. S. Ramakrishnan. 1981. Cropping and yields in agricultural systems of the north eastern hill region of India. *Agro-ecosystems* 7:11–25

Verma, L. R., R. P. Awasti, and A. K. Agnihotri. 2000. Horticulture development: productivity and sustainability imperatives. *Natural Resources and Development in Himalaya*, pp. 139–159. New Delhi: Malhotra Publishing House

CHAPTER 12

Conservation and Livelihood Links of Joint Management of Protected Area in Phutoei National Park, Thailand

Chandra Shekhar Silori

INTRODUCTION

The Joint Management in Protected Areas (JoMPA), a participatory tool of managing protected areas (PA), was introduced in Thailand in 2004 (Danida 2008). It aims to bring together different levels of actors and stakeholders involved in PA management to effectively promote a participatory management of PAs, establish a joint management of the PA model that can be replicated into other PAs, and enhance the capacities of the involved organizations and stakeholders in the process, besides focusing on improving livelihood opportunities for forest-dependent communities. Generally, in a typical JoMPA model in Thailand, the key stakeholders involved include authorities of protected areas, local communities, local authorities, civil society organizations, and the private sector. Support is provided to the Department of National Parks, Wildlife and Plant Protection under the Ministry of Natural Resources and Environment, which is, at present, the key government partner. Support is also given to a range of non-governmental organizations (NGOs) that are considered as important partners in maintaining the joint management activities. The introduction of JoMPA was necessitated by the continued loss of biodiversity, degradation of the ecosystem, loss of livelihood opportunities for the rural poor, and the lack of democratic involvement of a broad range of stakeholders in PA management (Chalerplap 2008).

Joint management in protected areas is under implementation in 11 PAs in Thailand, together covering approximately 18% of the total geographical area of the country. One of the fundamental features of JoMPA is clear demarcation of different use zones in and around the protected areas, including the area under human habitation, cultivation, and biodiversity conservation, and combining traditional and scientific knowledge in PA management. Especially the forest areas frequented by local communities to meet their livelihood needs are clearly demarcated and separated from the areas identified for biodiversity conservation. Such a practice not only helps avoid any potential conflict with the park authorities but also encourages neighbouring communities dependent on the forest to live peacefully and without causing damage to the surrounding forest. These initiatives are also supported by forming institutions at different levels—village level, national park jurisdiction, and province level. This case study was conducted in Phutoei National Park, in the Suphanburi Province in Western Thailand. The major objective of the case study was to understand the process of JoMPA and the roles of different actors and the resultant benefits in terms of the livelihoods of the dependent communities and the status of biodiversity conservation in the national park area.

STUDY AREA

Phutoei National Park (PNP) is one of the 17 protected areas of a large conservation network in Western Thailand known as Western Forest Complex (WEFCOM) conservation network (WCS 2010). PNP, located on the south-eastern fringe of WEFCOM, is surrounded by 22 villages, 7 of which are in the vicinity of 2 km from the national park boundary. One of these seven villages, Huai Hin Dam, which is located on the south-western fringe of the park, was selected for this case study. Huai Hin Dam has been developed as a model village to promote the concept of JoMPA in order to replicate the model in other neighbouring villages in similar situations. Therefore, research aspects in the village focused on not only understanding the conservation and livelihood linkages of JoMPA, but also the process of implementing this participatory management tool, including key stakeholders involved in the process and the roles played by them in making this a successful management initiative.

METHODOLOGY

The results presented in this chapter are based on a short-term study conducted in Huai Hin Dam Village. At the time of the case study, the village was inhabited by a total of 53 households. For this case study, 13 households (25%) were sampled using the stratified random sampling method, and through participatory, semi-structured questionnaire interviews. The questionnaires for participatory interviews were developed after rapid consultations and focused group discussions with different stakeholders, including community representatives, national park authorities, and a local NGO involved in facilitating the process of JoMPA. This process helped in assessing the local socio-economic situation and understanding the issues related to biodiversity conservation in and around the national park, including how the process of JoMPA evolved and was implemented and the roles played by the different stakeholders in the process. The draft questionnaires were tested before being used for formal interviews at the household level. Participatory surveys with respondents focused on quantifying household-level income patterns, sources of income, and a comparative estimate of income before and after JoMPA in order to assess the impacts and linkages of JoMPA with respect to the livelihoods of the local communities. Some of the indicators of livelihood status were the change in awareness towards conservation among the local villages, the status of conflict within the community and with the national park authorities over the use of forest and other natural resources, the participation of women in various activities including decision making, and the status of basic infrastructure such as road connectivity to the nearest market, drinking water, school, and health centres.

Since it was a short-term study, a qualitative, comparative assessment of biodiversity conservation in and around PNP was done by documenting the perception of local communities and other stakeholders on selected indicators. Some of the indicators included were the status of forest regeneration, frequency of sighting wild animals and birds, incidences of hunting, forest fires, and tree felling; availability of non-timber forest products (NTFPs) such as bamboo shoots and other plant products for making natural dyes, and other environmental services such as watershed conservation, soil erosion control, and agriculture productivity. Institutional linkages and relationships were considered as important criteria in the implementation process of JoMPA. Therefore, indicator-based perception documentation was also conducted with the respondents to capture the changes in relationship among the key stakeholders, including local community groups, women groups, national park authorities, and NGOs.

RESULTS

Demographic and Socio-economic Scenario

Huai Hin Dam Village is located in the Dan Chang District of Suphanburi Province. It is inhabited by the 200-year-old Thai Karen community. The Karen people migrated from Myanmar many centuries ago. After fleeing the war in Myanmar, way back in the Rattanakosin era, they were permitted to settle in the Kingdom of Thailand as Thai citizens. The average family size for the sampled population was nearly 5 (4.9) persons per household. The gender ratio was in favour of females, with 1133 females for every 1000 males. Of the total 13 households, 11 households reported agriculture as their main occupation, while daily wage labour and regular employment were the main occupations for the rest. The area of cultivation ranged from 2 to 50 rai (6.25 rai = 1 ha), averaging nearly 15 rai (2.4 ha) per household. Agricultural cultivation was dominated by corn (53%), upland rice (23%), fruits such as papaya, and a few vegetables (19%).

Collection of NTFPs and plant-based dye for making handicraft articles were subsidiary occupations for the villagers.

Household income patterns revealed wide disparity, ranging from 13 000 baht ($383)[1] to 318 000 baht ($9364) per annum, averaging 134 000 baht ($3946). Agriculture was the major source of household income, contributing about 63%, followed by 23% from regular employment in the government or private sector, 11% from the sale of natural dye-based handicraft articles, 2% from NTFPs, and the remaining 1% from daily wage labour (Figure 1). To further understand the income patterns of different socio-economic groups at the household level, sampled households were grouped under three broad categories, based on the land area cultivated by them, such as the marginal-to-small farmer cultivating 0–10 rai, the medium farmer cultivating 10–20 rai, and the large farmer cultivating more than 20 rai. The average annual income for these three categories was calculated to be 60 975 baht ($1795), 120 900 baht ($3560), and 318 333 baht ($9374), respectively. The results of the contribution by sources to the average household income for these three categories are presented in Figure 2. Across all the three categories of farmers, agriculture emerged as the major income contributing sector, while the ratio of the other sectors varied. For small farmers, besides employment, income from the forest sector in the form of NTFPs and from the sale of natural dye-based handicraft articles contributed

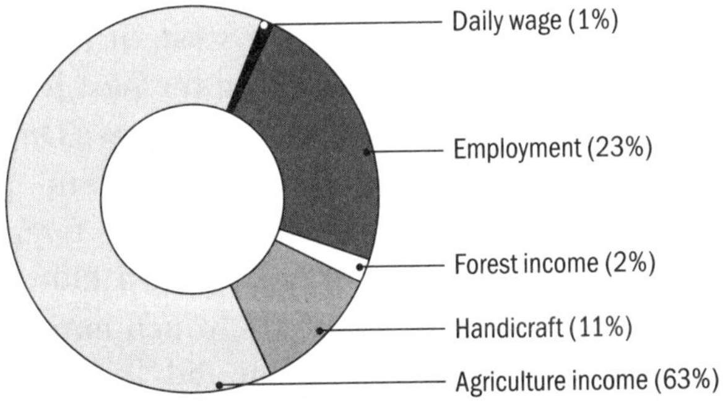

Figure 1 *Household income pattern in Huai Hin Dam Village in Suphanburi, Thailand*

[1] $1 = 33.96 Thai baht

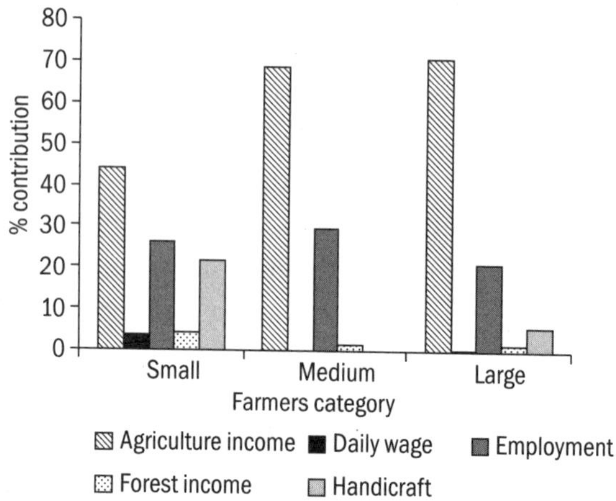

Figure 2 *Contribution to the household income by different sources*

nearly 26%, indicating the importance of the forest in sustaining their livelihoods. Comparatively, for medium to large farmers, the contribution from the forest sector was 1.6% and 7%, respectively. Interestingly, for the only landless household among the sampled population, of the total average income of 45 000 baht ($1325) per annum, 93% came from the sale of natural dye-based handicraft items and the remaining 7% from the sale of bamboo shoots.

Conservation and Livelihood Links

Forest quality and biodiversity values

The status of forests, in general, in the western region of Thailand has reported many ups and downs during the past few decades, and the Karen people have witnessed these changes. During the early 1970s and the 1980s, slash-and-burn agricultural practices and logging concessions were largely responsible for forest destruction. There was a ban on these practices in the late 1980s, supplemented by regeneration and conservation efforts, which initiated the process of recovering degraded landscapes, but did little to contribute to the biodiversity values, as most of such efforts were dominated by monoculture plantations. Further, the relocation of cultivated lands from such areas caused livelihood insecurity among the forest-dependent groups, including the Karen. Responding to such a situation, the Karen community of Huai Hin Dam Village formed a community forestry group in 1994 and initiated a forest regeneration and conservation programme on their own, using their traditional skills and knowledge of forest conservation.

The villagers classified the forest into six main types, based on their utilization. These are *Thikhikhwaju* (water catchment), *Thikhi* (watershed), *Chue-long-ong-khoe-ong* (for day-to-day use), *Chue-long-chai-soei-chai-wa* (for timber use), *Chue-long-ma-ong* (cultivation area), and *Sawakhu* (spiritual use) (ThCCP-RECOFTC 2008). However, these efforts suffered a setback with the notification of Phutoei National Park on 30 September 1998. Access to the park forest, including the cultivated lands which were now inside the notified area, was restricted, leading to conflicts between the park authorities and the villagers. Despite many efforts, these conflicts could not be resolved. However, with the introduction of JoMPA in 2004, Hui Hin Dam was chosen as one of the pilot villages, recognizing its past efforts in conserving and regenerating the degraded forest in and around the notified national park area.

Through a consultative and participatory approach, selected forest areas in and around PNP were managed, following a community forestry (CF) management plan. The CF plan reclassified the forest area into three main use patterns. These included 'conserved forest area' (consisting of *Thikhikhwaju*, *Thikhi*, and *Chue-long-ong-khoe-ong*), covering 10 800 rai, 'utilization forest' (*Chue-long-chai-soei-chai-wa*) covering 1800 rai, and 'controlled area', having rotational agriculture land (*Chue-long-ma-ong*), covering 1200 rai cultivated by the villagers under a common ownership. Such a clear demarcation of land was agreed on by both the parties, and private ownership of the rotational agricultural land and cultivation of the spiritual area or sacred forest such as mountain ridges, watershed areas, or areas near streams were not permitted. The rotational agricultural lands were kept outside the national park boundary. Logging and wildlife hunting were prohibited in the conservation forest, while timber harvesting from the utilization forest was regulated by the CF management committee through an approval process, limiting the number of harvestable trees to a maximum of 25 per year at a nominal fee of 100 baht/tree.

To report the changes in forest quality and biodiversity values after the implementation of JoMPA, respondents were asked to compare the current status of selected indicators against a 10-year-earlier situation, using a ranking method on a scale of 1 to 4 (poor, average, good, and very good). The ranking values given by the respondents for each indicator were then averaged. As presented in Figure 3, all the indicators have reported improvement during the last 10 years. While forest regeneration was rated as improving, hunting of wild animals and birds had reportedly declined. Similarly, the availability of NTFPs, including bamboo shoots, was rated as increasing. The response on the bamboo shoots was recorded separately, because it was one of the important NTFPs that fetched cash income to the local villagers. The villagers further added that a substantial decline in hunting resulted in increased diversity and population of wild animals and birds. While the number of wild boar (*Sus scrofa*), barking deer (*Muntiacus muntjak*), sambar deer (*Cervus unicolor*), rhesus macaque (*Macaca mulatta*), palm civet (*Paradoxurus hermaphroditus)*, slender loris (*Loris tardigradus*), common squirrel (*Callosciurus* spp.) and others

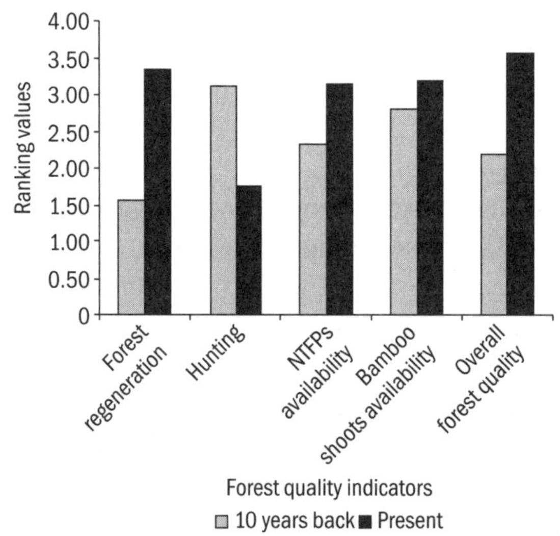

Figure 3 *Perception of respondents on the comparative assessment of conservation status in Phutoei National Park (PNP)*

have reportedly increased, bird species such as horn bills, peacocks, and jungle fowls have returned to the area. Some of these reports were also confirmed by the park authorities.

Forest quality and income patterns

The improved quality of the forests also yielded livelihood benefits to the forest-dependent villagers. The household interviews listed 8 edible and 15 medicinal plants collected by the villagers from the surrounding forests, besides fuelwood for cooking needs, and bamboo and timber for house construction. Although the cash contribution from the sale of NTFPs, mainly bamboo shoots, is about 2%, the edible herbs and other forest products are an essential part of their diet. Natural dye-based handicraft items are another source of forest-based income for the Karen people. They use plant-based colours to dye traditional dresses such as *pathums*, and shawls, scarves, shirts, handbags, and purses. During the questionnaire survey, the respondents said they extracted natural dyes from more than 20 plant species from different plant parts. 16 of such plant species were identified during a participatory field survey with the women in the surrounding forest (Table 1). Also, they reported that, during the last few years, the availability of such plants has improved in the surrounding forests and homestead areas, providing opportunities to increase production and income. Consequently, the respondents linked the increased income from the natural dye-based handicraft to the better forest conditions and also the institutional support provided by the saving group managed by a group of local women.

Out of the 13 sampled households, 8 (63%) were involved in natural dye-based handicraft to earn cash income. On an average, about 11% of a household's annual income is earned from handicraft items. Since 1999, these activities are organized through a saving-credit group of 20 women who make and sell a variety of natural dye-based handicraft items.[2] The data obtained from the records of this group from 2004 to 2007 reported a constant increase in the sale of handicraft items and income from them (Figure 4). The annual sale of handicraft items increased from 48 973 baht ($1442)

[2]Payong, personal communication, 29 January 2009

Table 1 *List of plants used to make natural dyes by the Karen people of Huai Hin Dam*

Local name	*Common name*	*Scientific name*	*Family*	*Plant part*	*Colour*
Fhang	Sappan	*Caesalpinia sappan* L.	Caesalpinioideae	Bark	Red
Padauk	Burma Padauk	*Pterocarpus macrocarpus* Kurz.	Leguminosae	Bark	Brown
Taowanlek		*Ventilago calyculata* Tul.	Rhamnaceae	Leaf	Yellow Green
Hom		*Strobilanthes cusia* (Nees) Kuntze.	Acanthaceae	Leaf	Blue
Taklum	Garuga	*Garuga pinnata* Roxb.	Burseraceae	Bark	Brown
Kratin	Wild tamarind	*Leucaena leucocephala* (Lamk.) de Wit.	Mimosoideae	Bark	Brown
Plao	Croton plant	*Croton roxburghii* N.P. Balakr.	Euphorbiaceae	Bark	Brown
Hookwang	Indian almond	*Terminalia catappa* Linn.	Combretaceae	Leaf	Yellow
Mamuang	Mango	*Mangifera indica* Linn.	Anacardiaceae	Leaf	Green
Peka		*Oroxylum indicum* (L.) Kurz.	Bignoniaceae	Leaf	Green
Konta		*Harrisonia perforata* Merr.	Simaroubaceae	Fruit	Black
Mapraw	Coconut	*Cocos nucifera* L.	Palmae	Bark	Brown
Sak	Teak	*Tectona grandis* L.f.	Labiatae	Bark	Brown
Yapa		*Morinda coreia* Ham.	Rubiaceae	Leaf	Green
Unchun	Blue pea	*Clitoria ternatea* L.	Leguminosae	Flower	Blue
Daokajay	Cosmos; Tickseed	*Cosmos sulphureus* Cav.	Compositae	Flower	Yellow

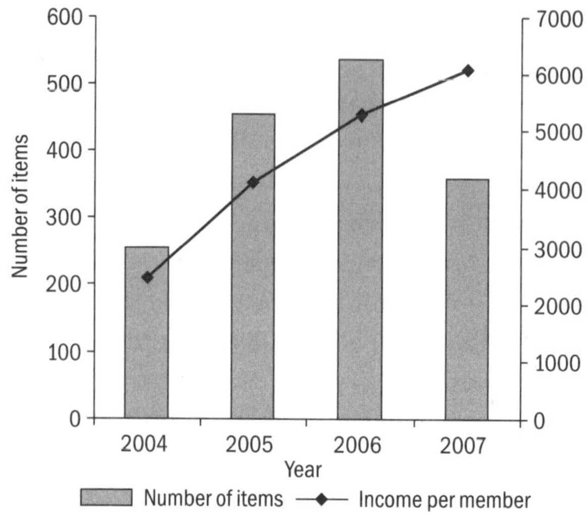

Figure 4 *Income generation patterns from the sale of plant-based dye handicraft items*

in 2004 to 121332 baht ($3572) in 2007, correspondingly resulting in the increase of the average annual income from 2249 baht ($66) to 6067 baht ($179) for each member.

Conservation awareness

Perception documentation on conservation awareness reported improvement among the local communities during JoMPA implementation phase. Though local communities had been using forest resources and accumulated traditional knowledge about conservation and the management of forests over the ages, the institutional approach to forest conservation under JoMPA helped them enrich their awareness. The results of the average ranking on a scale of 1 to 4 (poor to very good) regarding awareness levels among the local communities was rated 2.67 at the time of survey, compared to 1.88 about five years ago, indicating a contribution of streamlining conservation practices under the CF management plan and JoMPA. Among the younger generation, however, the awareness level averaged around 2.05 at present, whereas five years ago, it was 1.75.

Institutional building and linkages

Community forestry initiatives taken up by the villagers and later supported under the JoMPA scheme led to the development of village-level institutions. Besides a core group of community forestry management, a few other village-level groups, such as the women's group for natural dye-based handicraft items and a group of farmers supporting organic farming were promoted as a part of the institutional building process in the village. The existence and sustainability of such groups can be directly linked to the community forestry initiatives, which were further strengthened under JoMPA. The results of a perception study on the relationships of the local communities with the park authorities further prove this fact. The collective response on the relationships of the villagers with the national park authorities on a scale of 1 to 4 average 3.67 at present, compared to 1.5 about 5 years ago. The villagers almost unanimously perceived that their relationship with the park authorities had significantly improved. One of the key reasons for this, according to the villagers, was a change in perception among the park authorities towards the local communities. On the one hand, the park authorities recognized the forest-based livelihood needs of the villagers, while, on the other hand, they promoted regulated resource harvesting practices, particularly for bamboo shoots, bamboo, and timber. This practice also helped promote equal access and benefit sharing of the forest resources among the villagers, thus allowing natural regeneration to further increase the resource base. The results of the perception response on forest quality and biodiversity values, presented above, support these practices.

DISCUSSION

Over the past decade or so, several notable innovations have been made in the concept and practice of managing PAs. These innovations reflect the changing context for PAs and address emerging challenges, including those from climate change, forcing many traditional views to change, resulting in the advocacy of a new set of ideas and approaches for PA management. In this context, joint or collaborative management of protected areas has emerged as one of the

most practiced management strategies of PA recently (MacKinnon 2001). It has been tried out in many countries with varied degrees of success. There are a number of examples of joint management of PA from different parts of the world including the USA, Australia (De Lacy and Lawson 1997), Africa, and, more recently, from Asia (Oli 1999).

Co-management is a broad concept that encompasses a variety of ways in which stakeholders can jointly develop and implement a management partnership. It is particularly relevant in situations where active commitment and collaboration of stakeholders is necessary, and where access to the PA's natural resources is essential for local livelihood security and cultural survival (Borrini-Feyerabend 1996). The evolution and implementation of JoMPA in PNP is an example of such a collaboration. The traditional management practices, conservation values, and livelihood needs, on the one hand, and the willingness of the national park authorities to appreciate the livelihood concerns of the local communities, on the other hand, facilitated the collaboration and made it a success, as revealed from the results of the case study. It provides a learning process approach to implementation, contributes to increased trust and cooperation, and develops the people's sense of ownership by involving them in decision-making (Fisher 2000). Improved livelihood opportunities, better conservation status of the PA forest, and reduced conflict between local communities and the park authorities characterize the successful implementation of JoMPA in PNP.

As far as the livelihoods of the local communities are concerned, the contribution of forest resources, especially edible herbs and medicinal and dye-yielding plants, was substantial to the poor and marginal households. The absence of such resources meant spending more money to buy them or their alternatives from the market, which is difficult for the poor and marginal households due to their low income levels. Thus, such forest products meet the food and other needs of such households, and their relatively easy availability provides a safety net during hard times to poor households. Institutional building, promoting the concept of sustainable resource harvesting, equitable resource access, and improved awareness were other important features of JoMPA, which helped build a relationship of trust and confidence between the resource users and the park authorities. The prior experience and reputation of the Karen communities for conserving natural resources also contributed in accelerating the process and smooth implementation of collaborative PA management in the PNP.

CONCLUSION

Being implemented for the first time, the success of JoMPA in Thailand threw many challenges. Most of such challenges emerge due to ineffective participation of local communities in the planning and management of protected areas. Effective participation by local people is generally dependent on whether crucial issues such as livelihood security, security of tenure, land rights, usufruct rights, and equity are addressed by collaborative management agreements. In practice, achieving genuine participation in decision-making and benefit sharing has proved difficult. Therefore, action research is increasingly being advocated and used to overcome such challenges. Since some decision makers and policy makers may not be completely convinced by the concept of JoMPA, the success and replication of models like Huai Hin Dam may not prove easy. The lack of coordination between policy makers and field-level practitioners is another challenge that may hinder the outreach of JoMPA. It is also equally important to keep in mind that a co-management approach should not lead to the neglect of core conservation concerns. Co-management programs have to keep in mind the PA's management plans and legal frameworks; collaboration is a means of achieving management objectives and not an end in itself. Co-management is meaningless if the core values of the PA cannot be protected. Likewise, core values cannot be protected if the relationship with neighbouring communities is adversarial and fundamental issues such as their livelihood security are neglected (Phillips 2000; ICEM 2003).

REFERENCES

Borrini-Feyerabend, G. 1996. *Collaborative Management of Protected Areas: Tailoring the Approach to the Context.* Switzerland: International Union for Conservation of Nature-the World Conservation Union, 67 pp.

Chalerplap, S. 2008. *Achievement of Joint Management of Protected Area in Western Forest Complex Project (April 2004–March 2008)*. Bangkok, Thailand: SuebNakhasathien Foundation

Danida. 2008. Danida in Thailand. Details available at <www.thdkenv.info/eng/jompa.php>

De Lacy, T. and B. Lawson. 1997. The Uluru-Kakadu model: joint management of aboriginal-owned national parks in Australia. In *Conservation Through Cultural Survival: Indigenous Peoples and Protected Areas*, edited by S. Stevens, pp. 155–188. Washington, DC: Island Press

Fisher, R. J. 2000. Experiences, challenges and prospects for collaborative management of protected areas: an international perspective. Paper prepared for the symposium *Adaptive Collaborative Management of Protected Areas: Advancing the Potential*, organized by Cornell University, Ithaca, New York, 16–19 September 1998

ICEM (International Centre for Environmental Management). 2003. Regional report on protected areas and development. Review of protected areas and development in the lower Mekong River region, Queensland: ICEM

MacKinnon, K. 2001. Integrated conservation and development projects: can they work? *Parks* 11(2):1–5

Oli, K. P. (ed). 1999. *Collaborative Management of Protected Areas in the Asian Region*. Kathmandu: International Union for Conservation of Nature

Phillips, A. 2000. Benefits beyond boundaries: taking a broader view about the future of protected areas. [Executive Summary]. In *The World Commission on Protected Areas*, 1–2, edited by A. Galt, M. Vinton, and T. Sigaty. Pakse, Second Southeast Asia Regional Forum, PDR, vol. 2, International Union for Conservation of Nature Vientiane, Lao PDR

ThCCP-RECOFTC (Thailand Country Programme, Regional Community Forestry Training Center for Asia and the Pacific). 2008. Ban Huai Hin Dam community forestry. [Project Report]. Bangkok: ThCCP-RECOFTC

WCS (Wildlife Conservation Society). 2010. Best of the conservation: Wildlife Conservation Society and the western forest complex of Thailand, 18 pp. New York: WCS

CHAPTER 13

Sustainable Utilization and Management of Timur by Communities in Uttarakhand

Chandra Prakash Kala and Nehal Ahmad Farooquee

INTRODUCTION

The interaction and interrelationships of rural mountain people with their major economic activities such as agriculture, pastoralism, and other major natural resources is well understood and documented in a number of publications (Blaikie 1989; Ives and Messerli 1989; Dewalt 1994; Kala 1998a, 2007; Maikhuri, Semwal, Rao, *et al.* 1997; Farooquee 1998). This interaction is very delicate and varies in space and time according to the nature and extent of dependence on minor forest products. Much before the advent of pastoralism and agriculture, mountain forest dwellers had developed close links with the surrounding plant resources. However, the level of this interdependence is yet to be understood in quantitative terms. It is well known that minor forest produce contributes towards a balanced diet in resource-poor communities who are generally located in remote and far flung areas. This aspect has not attracted the attention of researchers as yet. The real value of wild edible plants is more pronounced during periods of economic and environmental crisis and the economic return from such collections has been well documented in the Himalayan region and elsewhere (Farooquee and Saxena 1996; Farooquee 1998; Shackleton, Dzerefos, Shackelton, *et al.* 1998).

In the Himalayan traditional societies, the landless and poor people have adopted strategies to optimally utilize resources for domestic consumption and supplement their household income. During lean periods or during economic crisis, there is an increased reliance on wild plants by the entire household, adopting various strategies under different geographical and social conditions (Farooquee and Saxena 1996; Hoon 1996). It is also well established that the more uncertain the environment, the greater is the adoption of strategies to extend the availability of resources (Campbell 1987; Shackleton, Dzerefos, Shackelton, *et al.* 1998). In most societies, these strategies are regarded as compulsion, whereas in others they are so well integrated and built into their social and cultural fabric that they are referred to as traditions. For example, in the Bhotiya Transhumant pastoral communities of Pithoragarh and Chamoli districts in Uttarakhand, Himalaya, timur (*Zanthoxylum armatum* DC.) collection and its barter exchange within and outside their communities in and outside their region are socially and culturally acceptable. Being an important plant species, the present study was undertaken to (1) analyse the status of the local people's indigenous knowledge on different uses of timur, (2) determine the practices employed to extend the availability of this shrub, (3) assess the traditional mechanism developed on barter exchange, and (4) document the quantity of its collection at village level.

MATERIALS

Timur (*Z. armatum* DC.) belongs to the family Rutaceae. It is found in the warmer valleys of the Himalaya ranging between 1000 and 2100 m above mean sea level. It is also grown in the Eastern Ghats of Odissa and Andhra Pradesh at 1200 m and the lesser Himalayan range in north-eastern India (Naga Hills, Meghalaya, Mizoram, and Manipur). Hooker (1875) described 11 species of genus *Zanthoxylum* in India, of which six occur in the Himalayan region. In the Uttarakhand Himalaya (Kumaon and Garhwal), four species of *Zanthoxylum—Z. alatum* Roxb., *Z. acanthopodium* DC., *Z. oxyphyllum* Edgew., and

Z. budrunga—are reported. Several uses of *Z. armatum* have been recorded from among many different communities of Southeast Asia, from China to Sri Lanka. In the Indian system of medicine, it is used extensively for various ailments. The fruit, seed, and bark have deodorant, disinfectant, and antiseptic properties that recently found good use in soap and dental cosmetics (CSIR 1989).

STUDY AREA

The Himalayan state of Uttarakhand is located on the eastern side of Western Himalaya. It spans an area of 53 485 km^2 and is inhabited by a population of 8479 562 people, of which 78% fall under the rural category (Census of India 2001). The districts of Pithoragarh and Chamoli are situated on the north-eastern part of the state. This study is based on the fieldwork carried out in the Munsyari and Dharchula subdivisions of Pithoragarh district in Kumaon (29°27′ N to 30°49′ N and 79°50′ E to 81°3′ E) and in the Joshimath sub-division of Chamoli district, Garhwal (30°18′ N to 31°25′ N and 79°50′ E to 81°3′ E). The subdivisions of Dharchula and Munsyari are located in north-eastern Uttarakhand, bordering India, Nepal, and China, whereas Chamoli district is situated in the north-central part of the state.

Both Pithoragarh and Chamoli are well known for their rich biodiversity, ethnic communities, traditional resource use pattern, and indigenous knowledge systems. They are also known for the origin of rivers like River Gori-Ganga and River Kali-Ganga from Pithoragarh and River Alaknanda from the Chamoli. Pithoragarh and Chamoli are also famous for their rich and diverse culture, tradition, and mythology. The diverse social and economic setting of the region has been well documented by Pant (1935), Raipa (1974), Hoon (1996), and Joshi, Garia, Hasnain, *et al.* (1999). The vegetation of this region has been documented by Naithani (1984), and Kala (1998a, 1998b, 2002). According to Naithani (1984), about 1700 species of vascular plants have been reported in the Chamoli district, and many of these plant species have medicinal value.

The surveyed villages of Munsyari are located at an altitude of 1400–1800 m, while the villages of Chamoli are located at

1500–1800 m. The major social groups residing in this region are the Bhotiyas, who are demographically the dominant group; the other groups are Brahmins and Thakurs (upper caste Hindus) and the lower caste Harijans. The Harijans in Kumaon and Garhwal Himalaya are traditionally called *silpkar* (artisans), and include several occupational castes such a soil extractors, carpenters, tailors, and copper and iron workers whose caste hierarchy position, unlike Harijans elsewhere in India, is rooted in political and social bias rather than in the traditional Hindu belief of purity and pollution (Upadhyay 1990). The Bhotiyas and the Harijans are often found settled together in a single village setting, but all the three social categories are rarely concentrated in a single village. In fact, in the collection and barter exchange of timur, only the Bhotiyas and the Harijans were involved directly.

Due to the topography of this region, there is a vast altitudinal difference within a short distance or area. This has resulted in great climatic variations and, as a result, several micro-climatic zones. Even in a single slope down the gradient, there are vast differences in climate (temperature, humidity, snowfall, wind), and so on in vegetation and agricultural patterns. The higher elevations have relatively cooler climatic conditions and have different cultivable crops such as beans (*Phaseolus vulgaris*), mustard (*Brassica compestris*), and potatoes (*Solanum tuberosum*), whereas the lower elevation, which has warmer climatic conditions, is known for the cultivation of rice and wheat as the major crops besides citrus fruits such as oranges and lemons. Although, timur grows naturally as a common shrub in several places in Uttarakhand Himalaya, it thrives luxuriously in terms of plant density and intensity of fruiting in the comparatively warmer regions like the foothills and valleys all along the Gori-Ganga and Kali-Ganga rivers in Kumaon, and in the valleys of Joshimath in Garhwal.

METHODOLOGY

For this study, we conducted a field survey in the selected villages and in small and big market centres related to the collection and sale of timur in and outside Kumaon and Garhwal. Fieldwork was carried out using a structured questionnaire survey to collect quantitative data on the collection of timur at the household level. Samples for household survey were stratified across different sections of the village society (caste, sex, age, educational, and occupational categories). Eight villages representing the existing specificities such as village size, altitude, existing forest type and size, aspect, landholding, agriculture type, distance from road head and so on were surveyed to gather information on the indigenous uses and commercial collection of timur. All types of small and big local markets were also surveyed for the total quantity of timur being traded from the region.

The participant observation method was employed to understand the local methods and techniques involved in collecting and processing timur fruits. Local resource persons were interviewed separately to elicit specific knowledge of indigenous timur uses and methods. Qualitative information so gathered was verified by cross-examination with different sections of the society. To understand the availability of timur in the studied villages, a detailed phytosociological investigation was conducted in forests where timur was abundant. In this study, villages endowed with the presence of timur were studied. The density of timur and their major associated plant species were also recorded from the eight villages. To assess the availability of timur in the wild, 10 quadrats of 5 m × 5 m were laid around each of the eight villages, and the number of individuals along with other dominant tree and shrub species were enumerated in each quadrat.

RESULTS

Availability and Distribution Patterns of Timur

Timur is usually found growing in barren and less fertile regions in mountain valleys, as well as in the forest as an understorey shrub. It thrives because no herbivores browse on it since it is a thorny shrub, and it propagates easily by seeds. Until recently in the Kumaon and Garhwal regions of Uttarakhand Himalaya, timur was not planted because it is easily available and recognized as a hardy species. Apart from its usual propagation by seed, it can also be propagated through vegetative cutting (DenHertog and Wiersum 2000). Some of the other common uses are as fuelwood and for fencing. Timur plants growing naturally in cultivable land are not acceptable to the farmers, and except for one or two plants on the edge of crop fields, others are removed from this region. This could probably be one of the factors for timur growing in barren areas and wastelands rather than on fertile agricultural land. One or two plants are also protected in homestead land for small domestic requirements.

In the Kumaon region, timur was found to be growing well in the valleys of Gori-Ganga and Kali-Ganga. Those villages located between 1300 and 1800 m have been found with good timur forests, and the fruits from these villages are preferred for domestic purposes and for commercial collection. There are around 25 villages in the Gori-Ganga valley and 14 in the Kali-Ganga valley from where the commercial tapping of timur is done, primarily due to its availability and good medicinal value. In all these villages of Munsyari and Dharchula, timur plants mostly grow on the northern and north-eastern slopes and less on the southern and south-eastern slopes. It grows in groups and forms a community with *Ficus palmata*, *Principia utilis*, and *Pyricantha crenulata*. In the Garhwal region, the density of timur was highest around the Naurakh village (841 individuals/ha), followed by Nauti (366 individuals/ha) and Bhimtalla (260 individuals/ha). Besides *F. palmata*, *P. utilis*, and *P. crenulata*, 25 major woody species associated with timur were found in the study area.

Indigenous Knowledge and Uses of Timur

Timur is extensively used in the high-altitude societies of Uttarakhand Himalaya. The Bhotiya, a dominant ethnic community of this region, has traditionally evolved a number of indigenous uses of timur in their daily lives. It is interesting to note that the Bhotiya use timur in more number of ways than any other ethnic groups, and mostly use its fruit for different purposes. The major uses of timur in the Bhotiya food habit are in the form of condiment, spice, medicine, sauce, and pulse (CSIR 1989; Kala, Farooquee, and Dhar 2005; Kala 2010 a, b). In the preparation of certain traditional dishes of this region, timur is very popularly used as a flavouring agent or spice. As part of their adaptation in this cold region, the Bhotiya make soup from the dried fruits called *hag*, and the entire family eats it more frequently during winter to keep warm. Similarly, another recipe that is quite common and used round the year is a sauce kind of preparation locally called *dunkcha*, mostly eaten with *roti* (Indian bread).

Some of the major uses of timur as medicine include cures for several common ailments such as toothaches, common cold, cough, and fever. To cure toothache, a timur fruit (fresh or dry) is pressed over the affected tooth and kept until there is relief from the pain or the pungent property of the fruit is lost, after which a fresh one replaces it till the pain is relieved. The young shoot of the timur (fresh) is used as a brush stick for cleaning teeth, a practice similar to the use of Neem (*Azadirachta indica*) in the plains of India. People have also started using the powder made from the dry fruits of timur for cleaning teeth; this, however, is an imitation of *Ayurvedic* tooth powder sold in the Indian market (Kala, Farooquee, and Dhar 2005). During winter, the most common health complaint of this region is *bai-gola*, whose conspicuous symptoms include eructation and water coming through the mouth, stomach pain, swelling of the stomach, stiffness, vomiting, and acidity. This problem is cured mainly by consuming timur soup. Etiologically, this is considered to be caused by heavy physical labour under extreme cold conditions in winter. Patients suffering from this problem are given timur soup three or four times a day.

Timur is also used as vermicide and pesticide; a few dried fruits are kept along with the woollen garments to dispel worms (CSIR 1989; Mao 1993). The unripe fruits are crushed and pounded into a soft paste and applied on the lower limbs to repel leeches. The other popular use of timur in the Bhotiya community is as a stimulator in brewing liquor, where a few dry timur fruits are added to the substrate to be fermented. This practice, however, makes the liquor toxic, and only highly addicted people can take such alcohol.

Besides the uses of timur fruits, the thin branch of the timur is commonly used as a walking stick in the mountains as well as in the plains of India. It is a popular walking stick because it is hardy, looks elegant, and is also believed to have religious significance—magical power—to ward off evil influences or malevolent forces.

Timur was most commonly used to heal *bai-gola* for which all respondents responded in its favour. The next most common use was as a vermicide, followed by its use in making soups, curing fevers, coughs, and colds. On the other hand, a large number of respondents said that timur was no longer used in preparing alcohol, followed by its use as a spice and a leech repellent.

Collection, Processing, and Management of Timur

The best time for collecting the timur fruit is October, when the fruit ripens to a dark red, which indicates its maturity and readiness for harvesting. The collection of its fruit starts from the first week of October and continues till early November. Only ripened fruits contain the richly active properties for which timur is used. Despite the extensive indigenous uses of timur, it does not receive major attention from the people for its collection. Since, only a small quantity of timur is required to meet the local requirement, children and women are the primary collectors of its fruit and, in most cases, the collection is opportunistic, based on the priority basis. However, in the recent past, the commercial harvesting of the fruits has led to indiscriminate collection and disputes amongst various local collectors. As a matter of fact, village-level institutions had to intervene to regulate and monitor the collection approach to avoid conflict.

Timur fruits grow from separate flowering shoots of the mature stem of a two- to three-year-old plant. Hand plucking of the fruit is the only possible method of harvesting. However, due to the thorny branches and small size of the fruit, the collection is rather difficult. Though lopping of fruit-bearing branches is convenient and helps access the branches that are otherwise out of reach, this causes tremendous harm to the plant and, consequently, leads to reduced production in subsequent years. Sometimes, it takes as long as 2 to 3 years for the recovery of fruit production due to the traditional lopping practices. Lopping the branches that would normally grow up to 6 m tall causes the plant to remain as a small bush. The fruits are sorted into two or three categories depending on their quality and are sun dried for two to three days. When the fruits are desiccated, they automatically burst open and the solitary seed comes out. However, to enhance the process of seed removal, the fruits are beaten with the help of a small wooden club. Seeds, having no local use, are separated from the edible dried crust for domestic use, or both, and the seed and the fruit crusts are sold in the market.

Until timur began to be harvested commercially, there were no *Panchayat* (village council) regulations for the management of timur growing areas and timur collection from the community forests. Later, due to the increasing demand of its fruit in the Indian pharmaceutical industries, the prices went up three fold, after which the attention of the villagers and the village council was drawn towards the management of timur grown in the community forests. Prior to commercial tapping, timur was sold at ₹12–15 per kg in the timur growing villages and local markets, whereas in 2000, its price in the local markets was ₹45 per kg and in 2005, it increased to ₹60 per kg. As a result of increasing market price and greater demand for timur in the local markets, there is rampant collection of timur particularly in the Munsyari and Dharchula regions of Kumaon. Following this, the *Sarpanch* (headman of the forest council) of most of the timur-rich villages is busy devising mechanisms for future regulations, management, and equitable sharing of the timur yield. In this regard, some of the common steps already enforced in some of the areas are restrictions to outsiders from collecting timur from their community forest; restriction to the villagers from collection before it is declared open, and banning (rather asking people to restrain from) wanton destruction while collecting timur fruits.

Social and Economic Dimensions of Timur Collection

The extensive utilization and consumption of timur in the Bhotiya communities irrespective of their settlement and location is still very common. It is required intensively during winter for a number of small uses within and outside the community. Over the years, the community has evolved a social mechanism for exchange and distribution between the villages endowed with timur and those that are not. The Bhotiya families residing in those areas where timur does not grow or is not available get their requirements of timur from relatives and friends living in timur-growing villages. Such reciprocity does exist even with those people who have moved out of the mountains and are living in far off towns and cities.

In the Bhotiya villages of Kumaon, about 14% of the total timur collected was for domestic consumption. Around 60% of its collection was sold to the village middleman, and rest was used for barter exchange and gifting to relatives and friends. In the Himalayan society, reciprocal exchange or barter of grains and crops was permissible for only a few items with the Harijan who are mostly blacksmiths by their traditional calling. The Harijan maintained the client relationships in villages far and near, which they visited annually for various artisan works. On their trips, they took the local forest products as intentional gifts for their patrons. This gesture was reciprocated by the patrons in kind in the form of vegetables, cereals, and other products. Despite no requirement for blacksmiths in the present time period, many of these poor people, who are mostly landless, still continue to visit their old clients, as well as those with whom they have established relationships in recent times. Through this relationship channel, timur is still distributed over several villages far and wide where it does not grow, and the poorest section of the society thus gets some economic return.

Timur has become a profitable non-timber forest product (NTFP) in recent years. The market value of timur has considerably gone up owing to the increase in demand. Even though seeds are not used locally, the collection of timur for seeds alone was also done, and accounted for about 50% of the total timur collection. In 2000, the market price of timur was about ₹45 per kg (dry weight including seed) in the collecting villages, whereas the price in the bigger markets in the plains was upto as much as ₹150–200 per kg. As a result, the villagers in the timur growing areas have started taking active participation in the collection of timur fruits. Now, even men have taken to timur collection, a job traditionally done by children and women. In 2000, the men constituted around 36% of the total people involved in timur collection.

Most of the people devoted their prime time in timur collection for cash; many started collecting timur well before maturity, as early as the first week of September, a month in advance. The maximum amount of timur was collected in Talla Dumor, followed by Malla

Dumor in Kumaon, whereas in Garhwal, the number of families involved in collection and the total quantity collected per village were much less. On an average, in 2000, a household in Munsyari and Dharchula earned around ₹700 in one season from a single NTFP of the area. It has been estimated that a person earned an average of over ₹50 per day, spending just four hours in collection, compared to the minimum wage of ₹77 per day for eight hours of work. However, since the collectors were mostly women, children, and old men who were otherwise economically unproductive, this earning from timur collection was profitable. The cash flow to timur-growing villages was also significant; for instance, the Talla and Malla Dumor together earned around ₹82 620 in 2000.

Conservation and Management

Since timur is not a fast-growing species and it possesses a low population size, the proliferation of woody weeds such as Lantana in the timur-growing areas are creating problems for the survival of this native species. Moreover, the system and mechanism of collection adopted by the local people is very harmful to its population size. In the Garhwal region, people harvest the entire plant during April–October to make walking sticks, which are in great demand in the local markets and religious places. Harvesting the entire plant before its flowering can adversely affect the regeneration and multiplication of the species (Kala 2010a). This process of unsustainable harvesting must be stopped immediately. The cutting of timur for fencing cultivated land is also destroying its population. Earlier, it was collected in low quantities for personal use but now the demand for timur walking sticks has increased due to the increase in the number of tourists and pilgrims in the Garhwal region; the commercial demand for timur fruit is just a two-year-old phenomenon in Kumaon. With the construction of road networks, the tourist and pilgrim inflow has increased manifold, and thus the increased demand for timur.

The plant association of timur is very interesting; normally it grows along with many other economically important species such as *Berberis aristata*, *Grevia optiva*, *Milia azadiracta*, *Woodfordia fruticosa*, *Pyrus pashia*, *Rubus ellipticus*, *Rhus parviflora*, and *Celtis australis*. *Berberis* sp. also has very high economic potential as pharmaceutical companies use its root to prepare medicine. The conservation of these economically important plant species is important as the local and commercial interest converges on these plants. Unfortunately, the traditional values and practices have been changing rapidly and the interpersonal relationship developed and nurtured between these people and the forces of commercialization ignore the natural resources and rules. The people who were once custodians of these vital resources are now rapidly harvesting these plants.

DISCUSSION

The high rate of destruction of timur plants and the overexploitation of its fruit are major impediments to conservation. This has posed a serious threat to the advancement of ethnic and rural cultures and the economy of the people. Although indigenous and customary resource management practices had earlier placed valuable restraints on the destructive exploitation of timur, due to the newly acquired economic value, the same people who once used them judiciously became the ruthless and destructive collectors. The crux of the argument is: can the timur shrubs be conserved when the commercial race for its fruit is at its peak and maximization of income is the main concern of the commercial timur harvest, especially when tapping of fruits is a labour-intensive exercise. It is also a well-known fact that in Kumaon and Garhwal Himalaya, there is a growing dependence on off-farm activities and more and more villagers are now involved in seasonal labour (Rawat, Farooquee, and Joshi 1996). The availability of labour, and the lean agricultural activity in October and November has influenced the gradual intensification of timur production. Some timur owners who lack sufficient family labour are also found to be making sharecropping arrangements with the marginal farmers to collect timur on their lands. Such arrangements for timur collection are an attractive alternative of off-farm employment to marginal farmers.

Following the increase in the market price of timur, the local people are not only involved in extensive lopping of the timur branches, but have also started collecting well before the appropriate time of its collection in competition with fellow villagers to get the maximum share. In the process, they have even undermined the profit that they would have got in the next season, and ignored the traditional land tenure methods that applied to the common property resource management. The management of common resources was enforced by traditional beliefs, which, even though being eroded, have been kept alive in the name of their traditional cultures. Such resources were conserved through open access user rights held by the entire village. Earlier, the natural ecosystems were managed at the village level through traditional land tenure regimes. Seeing the current destructive methods of timur collection, the forest *Panchayats* in some of the timur-rich villages are in the process of declaring policies prohibiting people from such methods of collection, and plan to introduce several regulatory measures.

Non-timber forest products (NTFPs) are always a blessing for the poor and underprivileged people whereas plantation of valuable plants for commercial purposes sooner or later becomes the monopoly of the economically powerful groups. Timur could be one of the species encouraged for plantation as a viable exchequer for resource-poor villagers. It can also grow on marginal and unproductive land as an understorey shrub without affecting the forest composition. Timur also has been reported to grow in the pine forest, where no other species usually grows. The total area under pine forest in Uttarakhand is very large, and could be utilized for timur cultivation/propagation. A proper agronomic study would give the exact economy of growing timur as an understorey shrub in some parts of Kumaon in India and Nepal, where timur is nurtured on the boundary of cultivable land.

Cultivation of timur as well as other NTFP of pharmaceutical importance could be highly profitable in the near future when the demand for such products is predicted to increase. The requirement for such resources in India is so vast that Indian markets have to import a huge quantity of NTFP every year mainly in the form of medicinal and aromatic plants (MAP) from Nepal. It has been estimated that more than 90% of the total NTFP harvested in Nepal is exported to Indian markets by various ways and means. According to Olsen (1998), 3300 kg of timur was transported to India in 1994–95 from one district (Gorkha district) of Nepal, whereas in Kumaon, in just one village of Munsyari, about 2000 kg of timur was collected 2000. Thus, the potential of timur production in the Uttarakhand Himalaya is great; the question is of proper tapping and networking. The timur harvest could be substantially improved by way of its plantation and better management practices.

CONCLUSION

Within Uttarakhand, various social groups have adopted different strategies of dependence on the timur plant. The Bhotiyas of Pithoragarh adopted the practice of collecting timur fruits, while the Bhotiyas of Chamoli developed the collection and trade of timur sticks to pilgrims visiting the shrines of Badrinath, Kedarnath, Gangotri, and Yamunotri every year. These people have evolved and established a mechanism for providing two different types of timur distribution in the region. As a result, they neither competed with each other, nor put too much pressure on the available resources that would hinder the natural timur propagation. Moreover, at the same time, they continued the barter exchange of timur with each other. The relationship between these two communities still exists, but in recent years, some pharmaceutical companies have started buying the timur fruit in bulk from this region, leading to competitive commercial activity among the villagers. To reduce the existing pressure on natural population of timur, attempts should be made to cultivate this species as a viable exchequer for resource-poor villagers. If concerted efforts are made, it can be grown on marginal and unproductive lands as an understorey shrub without affecting the forest composition.

ACKNOWLEDGEMENT

The authors would like to thank Jagat Singh Martoliya and Dinesh Nikhurpa for their help during field surveys.

REFERENCES

Blaikie, P. 1989. Explanation and policy in land degradation and rehabilitation for developing countries. *Land Degradation and Rehabilitation* 1:23–37

Campbell, B. M. 1987. The use of wild fruits in Zimbabwe. *Economic Botany* 41:375–385

Census of India. 2001. *Census of India 2001*. New Delhi: Office of the Registrar General, Government of India

CSIR (Council of Scientific and Industrial Research). 1989. *The Wealth of India: Raw Materials*, vol. XI. New Delhi: Publications and Information Directorate, CSIR

DenHertog, W. H., and K. F. Wiersum. 2000. (*Zanthoxylum armatum*) production in Nepal: dynamics in non-timber forest resource management. *Mountain Research and Development* 20(2):136–145

Dewalt, B. R. 1994. Using indigenous knowledge to improve agriculture and natural resource management. *Human Organization* 5:123–131

Farooquee, N. A. 1998. Development and the eradication of traditional resource use practice in the Central Himalayan transhumant pastoral society. *International Journal of Sustainable Development and World Ecology* 5:43–50

Farooquee, N. A., and K. G. Saxena. 1996. Conservation and utilization of medicinal plants in high hills of the Central Himalaya. *Environmental Conservation* 23(1):75–80

Hooker, J. D. 1875. *Flora of British India*, vol. 1. London: L Reeve & Co

Hoon, V. 1996. *Living on the Move: Bhotiyas of Kumaon Himalaya*. New Delhi: Sage Publications

Ives, J. D., and B. Messerli. 1989. *The Himalayan Dilemma: Reconciling Environment and Development*. London: Routledge

Joshi, A., P. S. Garia, and I. Hasnain. 1999. *Uttarakhand Update: A Statistical Profile for Development Planning*. New Delhi: Kanishka Publishers

Kala, C. P. 1998a. Ecology and conservation of alpine meadows in the Valley of Flowers National Park, Garhwal Himalaya. [PhD thesis submitted to the Forest Research Institute Deemed University, Dehradun]

Kala, C. P. 1998b. Ethnobotanical survey and propagation of medicinal herbs for small farmers in the buffer zone of the Valley of Flowers National Park, Garhwal Himalaya. [Report submitted to ICIMOD, Nepal]

Kala, C. P. 2002. *Medicinal Plants of Indian Trans-Himalaya*. Dehradun: Bishan Singh Mahendra Pal Singh

Kala, C. P. 2007. Local preferences of ethnobotanical species in the Indian Himalaya: implications for environmental conservation. *Current Science* 93(12):1828–1834

Kala, C. P. 2010a. Assessment of availability and patterns in collection of Timroo (*Zanthoxylum armatum* DC.): a case study of Uttarakhand Himalaya. *Medicinal Plants* 2 (2):91–96

Kala, C. P. 2010b. *Medicinal Plants of Uttarakhand: Diversity Livelihood and Conservation*. Delhi: Biotech Books

Kala, C. P., N. A. Farooquee, and U. Dhar. 2005. Traditional uses and conservation of Timur (*Zanthoxylum armatum* DC.) through social institutions in Uttarakhand Himalaya, India. *Conservation and Society* 3(1):224–230

Maikhuri, R. K., R. L. Semwal, K. S. Rao, and K. G. Saxena. 1997. Agroforestry for rehabilitation of degraded community lands: a case study in Garhwal Himalaya. *International Tree Crops Journal* 9:89–99

Mao, A. A. 1993. A preliminary report on the folklore botany of Mao Nagas of Manipur (India). *Ethnobotany* 5:143–47

Naithani, B. D. 1984. *Flora of Chamoli*, vol. I–II. Dehradun: Botanical Survey of India

Olsen, C. S. 1998. The trade in medicinal and aromatic plants from central Nepal to northern India. *Economic Botany* 52(3):279–292

Pant, S. D. 1935. *The Social Economy of the Himalaya*. London: George Allen and Unwin

Raipa, R. S. 1974. *Shauka Simavarti Janjati* (in Hindi). Dharchula: RA Brothers

Rawat, D. S., N. A. Farooquee, and R. Joshi. 1996. Towards sustainable land-use in the hills of Central Himalaya, India. *International Journal of Sustainable Development and World Ecology* 3:57–65

Shackleton, S. E., C. M. Dzerefos, C. M. Shackelton, and F. R. Mathabela. 1998. Use and trading of wild edible herbs in the central Lowveld Savanna region, South Africa. *Economic Botany* 52(3):251–259

Upadhyay, H. C. 1990. *Harijans of Himalaya*. Nainital: Gyanodaya Prakashan

CHAPTER 14

Ethno-taxonomical Study of Medicinal Plants in the Nanda Devi Biosphere Reserve

C S Rana, J K Tiwari, L R Dangwal, and R D Gaur

INTRODUCTION

Ethnobotany has attracted much attention not only due to its many economic applications, but also because of its great academic and historical importance. It is a very broad field, including many aspects of botany and other disciplines (Tiwari 1986). The importance of plants for medicines and other use values has been well documented in the past (Badoni and Badoni 2001; Bhandari, Gaur, and Rawat 2001; Rana 2007). Primitive communities and tribes who live in the vicinity of forests, being close to nature, posses a deep, practical knowledge on indigenous flora pertaining to curatives, culture, religious beliefs, and myths (Maikhuri, Nautiyal, Rao, *et al.* 1997, 1998; Dangwal, Rana, and Sharma 2010). The Indian Himalayan Region (IHR) is a treasure trove of ethnobotany, especially of medicinal plants and related traditional knowledge. This region supports about 18 440 plant species, of which 1748 species have medicinal value and 675 species have edible properties (Negi and Gaur 1994).

Efforts have been made to conserve this rich biodiversity by establishing a Protected Area Network in the region. At present, there are 5 biosphere reserves, 28 national parks, and 98 wildlife sanctuaries in the IHR covering 51 899.238 km^2 (Kala 2005), distributed over tropical, temperate, and alpine ecosystems. Among these, Nanda Devi Biosphere Reserve (NDBR) is a prominent one, situated in Western Himalaya of Uttarakhand. Biodiversity rich, NDBR and the Valley of Flowers National Park form the core zone of the reserve. Rich cultural diversity adds to the biological diversity, making it one of the unique ecosystems in the region (Joshi and Samant 2003; Kala 2002). It has an area of 624.60 km^2 and an average altitude exceeding 4500 m above sea level, surrounded by high mountain ridges and peaks on all sides except on the western side, which features a deep and virtually inaccessible gorge (Maikhuri, Nautiyal, Rao, *et al.* 1997; Dangwal, Rana, and Sharma 2010).

Taxonomical studies in the IHR have brought out a number of valuable and informative publications, starting from the 18th century (Burkill 1965) and followed by many new insights by a number of researchers (Rau 1975; Naithani 1984, 1985; Deva and Naithani 1987; Hajra and Balodi 1995; Murti 2001; Bhatt, Bisht, and Gaur 1988; Dangwal, Kumar, Rana, *et al.* 2010; Gaur 1999; Gaur and Tiwari 1987; Gaur, Semwal, and Tiwari 1984; Nautiyal 1981; Negi and Pant 1992; Gaur, Negi, Pant, *et al.* 1993; Ansari and Ghananand 1985; Badoni 1988, 1989, 1990; Bhatt and Gaur 1992; Joshi and Pande 1995, 1997; Maheshwari and Singh 1984; Mumgain and Rao 1990; Pangtey, Rawat, and Samant 1989; Chaudhari, Sharma, Singh, *et al.* 2007). These studies have significantly added to the knowledge on the numerous useful plants that were not as well known earlier. Despite such extensive documentation on ethnobotany, there is little information available on ethno-taxonomy. In this study, therefore, we have focused on documenting ethno-taxonomy of the high-altitude medicinal plants of NDBR, applying the integrated approach of classical taxonomy accompanied by interviews of local inhabitants, *vaidyas*, priests, and other medicine men. The key objective of adopting this approach was to bridge the gap between the folk medicine practitioner and the contemporary natural therapists and researchers of natural product chemistry.

METHODOLOGY

Plant Collection and Specimen Preparation

Plant specimens were periodically collected from different ecosystems within Nanda Devi Biosphere Reserve, including alpine and sub-alpine regions. The monthly collection of plants was done every year from March to November between 2005 and 2010, in an altitude range from 1500 to 4500 m above sea level. Herbariums of collected species were prepared, following techniques suggested by Jain and Rao (1997) and deposited in the herbaria of the HNB Garhwal University, Srinagar Garhwal, Uttarakhand.

Folklore Survey

During the study period, ethnobotanical surveys were conducted in the tribal inhabitant areas of NDBR. The most frequently visited villages were Niti, Malari, Bampa, Gamsali, Garpak, Jelam, Jumma, Dronagiri, Kaga, Kosa, Phagti, Tolma, Lata, Peng, Suraithoda, Reni, Ringi, Subhain, Karchon, Tugasi, Raigari, Langsi, Tangari, Helung, Pagna, Salure, Peni, Adamath, Dungra, Selung, Palara, Gulabkoti, Urgam, Jakhola, Kamana, Dweng, Ganai, and Molta. The ethno-medicinal data was gathered from the tribal and local men and women *vaidyas* and from other knowledgeable informants using participatory interviews. The data on the names of plants, their parts used, their mode of processing and preparation, dosage, and various recipes and remedies were also gathered during such interviews. An exploratory visit was organized along with the informants into the nearby forests to collect herbarium specimens. Special efforts were made to build a rapport with the respondents, as some of them were not willing to disclose traditional knowledge on the use of certain plants. We spent sufficient time in the study villages to win the confidence of the villagers, respecting their customs and rituals, and being part of it. In order to authenticate the information, we validated the data using the triangulation method such as interviewing a number of respondents to get information on the use of the same plant species from neighbouring villages.

Enumeration

The taxonomical arrangement of the collected specimens followed Bentham and Hooker's system of classification (1862–83), except in a few cases where Hutchinson's concept (1973) of splitting certain families has been followed. The taxonomical categories, including genera and species within the family are mentioned along with the usual citation that are the vernacular names. The habit, habitat, locality of collection, flowering and fruiting periods, distribution status, and altitude ranges are described along with the ethno-medicinal properties. Finally, the rarity status of the species is mentioned based on the available information.

RESULTS AND DISCUSSION

A total of 50 plant species used as medicine are recorded during the present survey in the sub-alpine and alpine regions of the Nanda Devi Biosphere Reserve. These are medicinal plants frequently used by the communities, which we understood from our discussions with the local and traditional herbal healers. The detailed ethno-taxonomical information gathered is presented in Table 1. Of these 50 medicinal plants, 20 species fall under different threatened categories (for example, five species each in the 'endangered' and 'rare' categories, four species each in the 'vulnerable' and 'critically endangered' categories, and two in the 'near threatened' category).

This study has yielded some promising results about the use of medicinal plants, including unknown or lesser known uses of certain plants. It is hoped that the detailed observation on the traditional uses of plants in medicine by the tribal communities and primitive societies of this region will result in strengthening our knowledge on the Indian System of Medicine (ISM). A total of 50 important, threatened medicinal plant species belonging to 45 genera were documented during the study. Some of the following plants have the potential to be used as emergency drugs for primary health care: *Aconitum balfourii* for arthritis, *Aconitum heterophyllum* for chronic fever, *Astragalus candoleanus* for abdominal complaints, *Arnebia benthamii* for mental disorders, *Berberis aristata* for diabetes, *Podophyllum hexandrum* for leucorrhoea, *Potentilla fulgens* for

Table 1 *Medicinal plant species used by the local people of the Nanda Devi Biosphere Reserve*

Botanical name	*Local name*	*Family*	*Habit*	*Habitat*	*Locality*	*Altitude (m)*	*Flowering and fruiting*	*Part used*	*Form*	*Uses*	*Status*
Abies spectabilis (D. Don) Mirbel	Ragu	Pinaceae	Tree	Temperate	Auli	2400–3000	Mar–Nov	Bark, leaf	Paste, powder	For Cold, cough, and bronchitis	—
Aconitum balfourii Stapf	Meetha bish	Ranun-culaceae	Herb	Alpine	Kuari Pass	3000–4000	Aug–Nov	Tuber	Paste, powder	As an antidote of snakebite and scorpion sting: for rheumatic arthritis; as tonic.	VU
Aconitum heterophyllum Wall. ex Royle	Atish, Ateesh	Ranun-culaceae	Herb	Alpine	Gorson	3000–4000	Jul–Sep	Tuber	Paste, powder, decoction	For chronic fever, digestive disorders, diarrhoea, cold efficacy, diabetes, leucorrhoea; and as carminative.	CR EN
Allium stracheyi Baker	Ladmu faran	Liliaceae	Herb	Alpine	Neelkant	3000–4000	Aug–Oct	Leaf	Powder	As carminative; for cold, cough; and as a condiment.	VU
Angelica glauca Edgew	Choru	Apiaceae	Herb	Alpine	Dalisera	3000–3500	Aug–Sep	Root	Powder, decoction	For constipation, dyspepsia, and bronchitis.	EN
Arnebia benthamii (Wall. ex G. Don I.M. Johnston	Laljadi	Borag-inaceae	Herb	Alpine	Kuari Pass, Dalisera	3000–4500	Jun–Nov	Root	Paste, powder	As an antiseptic, hair tonic; for leucorrhoea, piles, diabetes, mental disorders; and as a tonic general weakness.	CR EN

Contd...

Table 1 *Contd...*

Botanical name	*Local name*	*Family*	*Habit*	*Habitat*	*Locality*	*Altitude (m)*	*Flowering and fruiting*	*Part used*	*Form*	*Uses*	*Status*
Asparagus filicinus Buch. Ham	Jhiran	Liliaceae	Shrub	Temperate	Merug, near Joshimath	2000–2700	May–Sep	Root	Powder, decoction	As a stimulant, tonic; for leucorrhoea, diabetes; to improve lactation; and in urinary complaints.	—
Astragalus condoleanus Royle ex Benth.	Rudra-vanti	Fabaceae	Herb	Alpine	Badari-nath	3000–4000	Sep–Oct	Root	Powder, decoction	For abdominal complaints, leucorrhoea; aphrodisiac, tonic in general weakness, for menstruation disorder, and diabetes.	—
Berberis aristata DC.	Kirmode	Berberi-daceae	Shrub	Temperate	Badagaon	1800–2700	Mar–May	Root	Powder, decoction	For urinary complaints, as a blood purifier, in ophthalmic infections, and diabetes.	—
Bergenia stracheyi (Hk. f. et. Thom.) Engl	Silpadi	Saxifra-gaceae	Herb	Alpine	Dalisera	3000–4500	May–Jul	Root	Paste, powder	As hair tonic, for cough, cold, abdominal ailments, liver complaints, asthma, piles, and urinary complaints.	Rare
Betula utilis D. Don	Bhoj	Betul-aceae	Tree	Alpine	Dalisera	3500–4000	May–Oct	Bark	Paste, powder	For skin ailments, plaster for dislocated joints, and in joint pain.	EN

Contd...

Table 1 *Contd...*

Botanical name	*Local name*	*Family*	*Habit*	*Habitat*	*Locality*	*Altitude (m)*	*Flowering and fruiting*	*Part used*	*Form*	*Uses*	*Status*
Calanthe tricarinata Lindley	Syoru	Orchidaceae	Herb	Temperate	Lamri, Auli	2500–3000	Mar–Sep	Root, leaf	Paste, powder	For eczema, sores, internal injury, and aphrodisiac.	—
Cedrus deodara (Roxb. ex D. Don) G. Don	Diwar	Pinaceae	Tree	Temperate	Niti valley	2500–3500	Sep–Oct	Oil	Decoction	For piles, arthritic pain, ulcers, skin ailments, fever, and dysentery.	—
Dactylorhiza hatagirea (D. Don) Soo	Hattajadi	Orchidaceae	Herb	Alpine	Valley of Flowers	3000–3500	July–Aug	Root	Paste, powder	As tonic, aphrodisiac; for diarrhoea, dysentery, and chronic fever.	—
Delphinium vestitum Royle	Nirbishi	Ranunculaceae	Herb	Alpine	Kuari Pass	3200–3500	Aug–Sep	Root	Paste, powder	For wounds, cuts and blisters; as antidote to snakebites and scorpion stings.	—
Dioscorea deltoidea Wall. ex Grisebach	Genthi	Dioscoreaceae	Climber	Temperate	Badagon, Joshimath	2000–2500	May–Oct	Root	Paste, powder	For ophthalmic disorders, cuts, wounds, ulcers, antiseptic; and to treat urinogenital disorders.	—
Ephedra gerardiana Wall. ex Stapf	Somlata	Ephedraceae	Shrub	Sub-alpine	Badrinath	3500–4500	Jul–Sep	Stem	Powder	For asthma, bronchitis, and cardiac complaints.	EN

Contd...

Table 1 *Contd...*

Botanical name	*Local name*	*Family*	*Habit*	*Habitat*	*Locality*	*Altitude (m)*	*Flowering and fruiting*	*Part used*	*Form*	*Uses*	*Status*
Fritillaria roylei Hk.	Kshir, Kakoli	Liliaceae	Herb	Alpine	Valley of Flowers	3000–4000	Jun–Oct	Root	Powder	As tonic in general weakness; for bronchitis, asthma; and to increase immunity.	VU
Geranium wallichianum D. Don ex Sweet	Laljari	Geran-iaceae	Herb	Alpine	Unile near Auli	2500–3000	Sep–Oct	Root	Sap	For earache, eye ailments, stomach disorders; and to improve eye vision.	—
Goodyera repens (L.) R		Orchi-daceae	Herb	Alpine	Tali	2500–3500	Jul–Oct	Root, leaf	Paste, powder	For fever, syphilis, and as a blood purifier.	—
Habenaria pectinata (J.F. Smith) D. Don	Ridhi-sidhi	Orchid-aceae	Herb	Chormy near Badagaon	Chormi near Badagaon	1700–2200	Jul–Sep	Root, leaf	Paste, powder	Semniferous and as a tonic in general weakness.	—
Hedychium spicatum Buch.- Ham. ex J.E. Smith	Ban-haldu	Zingib-eraceae	Herb	Temperate	Joshimath	1500–2200	Jul–Oct	Root	Paste, powder	As stomachic, carminative, tonic, stimulant; for asthma, indigestion, and bronchitis.	—
Juniperus communis L.	Bhitaru	Cupres-saceae	Shrub	Sub-alpine	Kauri Pass	2500–3200	Sep–Oct	Root, leaf	Powder, paste	For leucorrhoea, skin ailments, and dysmenorrhoea.	—

Contd...

Table 1 *Contd...*

Botanical name	*Local name*	*Family*	*Habit*	*Habitat*	*Locality*	*Altitude (m)*	*Flowering and fruiting*	*Part used*	*Form*	*Uses*	*Status*
Jurinea dolomaiea Boiss	Guggle	Aster-aceae	Herb	Alpine	Dalisera	3500–4000	Jul–Oct	Root	Powder, decoction	For leucorrhoea, menstrual disorders, mental disorders, gout, and urinary complaints.	EN
Malaxis muscifera (Lindley) Kuntze	Jeevak	Orchi-daceae	Herb	Subalpine	Taali near Gorson	2500–3000	Jul–Sep	Root, leaf	Paste, powder	As tonic in general weakness.	NT
Meconopsis aculeata Royle		Papaver-aceae	Herb	Alpine	Kuari Paas, Dalisera	3000–4500	Jun–Oct	Root	Purified powder	For unconsciousness, as tonic, and in nervous irritation.	Rare
Meconopsis robusta Hk. f. & Thomson	Laxmana	Papaver-aceae	Herb	Alpine	Kuari Paas, Dalisera	3000–4500	Jun–Oct	Root	Purified powder	In sterility.	Rare
Nardostachys grandiflora DC.	Maasi	Valeria-naceae	Herb	Alpine	Kuanri Paas	3200–4000	Aug–Oct	Root	Powder, decoction	For arthritis, cardiac complaints, and as aphrodisiac piles, hysteria, epilepsy, bronchitis, respiratory complaints.	CR EN
Paeonia emodi Wall. ex Royle	Chandra	Paeonia-ceae	Herb	Temperate	Lamari	1800–3000	Mar–Jun	Leaf	Paste, powder, decoction	For colic, blood dysentery, diabetes, urinary complaints, hysteria and epilepsy.	—

Contd...

Table 1 *Contd...*

Botanical name	*Local name*	*Family*	*Habit*	*Habitat*	*Locality*	*Altitude (m)*	*Flowering and fruiting*	*Part used*	*Form*	*Uses*	*Status*
Paris polyphylla J.E.Smith	Satwa	Liliaceae	Herb	Sub-alpine	Lamri near Auli	2500–3000	May–Sep	Root	Paste, powder	Respiratory disorders, anthelmintic, tonic; and as cardiac complaints, and asthma.	—
Picrorrhiza scrophulariflora Pennell	Katuki	Scrophul-ariaceae	Herb	Alpine	Dalisera	3000–4500	Jun–Jul	Root, leaf	Paste, powder	For chronic in bronchial asthma, as liver stimulant, and stomachic.	CR EN
Pinus wallichiana A. B. Jackson	Chew	Pinaceae	Tree	Temperate	Joshimath	2000–3000	Apr–Sep	Bark, leaf	Paste, powder	For rheumatic pain, asthma, bronchitis, and sprain.	—
Podophyllum hexandrum Royle	Shon kakadi	Podophy-llaceae	Herb	Alpine	Kuari Pass	3000–4000	Apr–Oct	Root	Paste, powder	Bile complaints, carminative, leucorrhoea, chronic fever, and diabetes.	VU
Polygonatum cirrhifolium (Wall.) Royle	Teetu	Liliaceae	Herb	Temperate	Merug, Badagaon	2000–2500	May–Jul	Root	Paste, powder, decoction	For leucorrhoea, blood purifier, in piles, to improve lactation, and an aphrodisiac.	—
Potentilla fulgens Wall. ex Hk.	Bajard-anti	Rosaceae	Herb	Alpine	Dalisera	2500–3500	Jul–Oct	Root	Paste, powder	For toothache and as and as	—

Contd...

Table 1 *Contd...*

Botanical name	*Local name*	*Family*	*Habit*	*Habitat*	*Locality*	*Altitude (m)*	*Flowering and fruiting*	*Part used*	*Form*	*Uses*	*Status*
Prunus cerasoides D. Don	Panya	Rosaceae	Tree	Temperate	Panya near Badagaon	2000–2500	Feb–Mar	Bark, seed	Paste, powder, decoction	For Internal injury, arthritis, sprains, and chest pains.	—
Quercus leucotrichophora A. Camus	Banj	Fagaceae	Tree	Temperate	Lamri, Auli	2500–2700	May–Oct	Gum	Paste	As a diuretic, for gonorrhoea, and as a digestive.	—
Rheum australe D. Don	Dolu	Polygonaceae	Herb	Alpine	Kuari Pass, Dalisera	3000–4000	Jun–Aug	Root, leaf	Paste, powder, decoction	Internal injury, joint pain, leucorrhoea, abdominal complaints, and as a tonic in general weakness.	EN
Rheum moorcroftianum Royle	Archu	Polygonaceae	Herb	Alpine	Dalisera	3500–4500	Jun–Sep	Root	Paste, powder	Chronic bronchitis, asthma, abdominal disease, and as tonic in general weakness.	NT
Satyrium nepalense D. Don	Mooshali	Orchidaceae	Herb	Sub-alpine	Chormi near Badagaon	1500–2000	Jul–Oct	Root, leaf	Paste, powder	For diarrhoea, bowl complaints, gonorrhoea, gonorrhoea, spermatorrhoea, anti-colic, diabetes, and as an aphrodisiac.	—
Saussurea gossypiphora D. Don	Hiyun-kaun	Asteraceae	Herb	Alpine	Panger Chula	4000–4500	Sep–Oct	Root	Paste, powder	For piles, cardiac complaints, as antidote for snake bites and scorpion sttings.	Rare

Contd...

Table 1 *Contd...*

Botanical name	*Local name*	*Family*	*Habit*	*Habitat*	*Locality*	*Altitude (m)*	*Flowering and fruiting*	*Part used*	*Form*	*Uses*	*Status*
Saussurea lappa (Dcne.) Sch.-Bip	Kuth	Aster-aceae	Herb	Alpine	-	2500–3500	Aug–Sep	Root	Paste, powder, decoction	For asthma, bronchitis, eczema, ring worm, and as a tonic.	Rare
Skimmia anquetilia N.P. Taylor & Airy Shaw	Nairpatti	Rutaceae	Shrub		Lamri near Auli	1500–3000	Sep–Oct	Leaf	Paste, powder	For psoriasis, leucoderma, and cooling effect.	—
Spiranthes sinensis (Persoon) Ames	Garud Panja	Orchid-aceae	Herb	Sub-alpine	Auli	2000–2700	Jul–Sep	Root, leaf	Paste, powder	As anti-colic, for leucorrhoea, and diabetes.	—
Tanacetum nubigenum DC.	Doana	Aster-aceae	Herb	Alpine	Dhak-wani	3500–4500	Jul–Sep	Root, leaf	Paste, powder	For chronic fever, as a tonic in general weakness, and to increase immunity.	—
Tanacetum dolichophyllum (Kitamura) Kitamura	Guggl dhoop	Aster-aceae	Herb	Alpine	Dalisera	3000–4500	Jul–Sep	Root	Paste, powder	For cuts, wounds, chest pain, paralytic affliction, and in arthritis	—
Taxus baccata L.	Thuner	Taxaceae	Tree	Temperate	Auli	2500–2800	Apr–Oct	Bark, leaf	Paste, powder	Cough and cold, cardiac complaints, abdominal diseases, and as anti-helmentic.	—
Thalictrum foliolosum DC.	Pilijadi	Ranun-culaceae	Herb	Temperate	Tapovan	1500–2500	Aug–Oct	Root	Paste, powder	Fever, headache, as carminative, in piles, blood purifier, ophthalmic afflictions, and herpes zoster.	—

Contd...

Table 1 Contd...

Botanical name	*Local name*	*Family*	*Habit*	*Habitat*	*Locality*	*Altitude (m)*	*Flowering and fruiting*	*Part used*	*Form*	*Uses*	*Status*
Thymus linearis Benth.	Marchya ghass	Lamia-ceae	Herb	Alpine	Auli, Gorson	2700–3000	Jun–Oct	Leaf	Paste, powder	Eczema, psoriasis, bronchial asthma, abdominal disease, and to regulate menstruation.	—
Trichosanthes tricuspidata Lour	Aladu	Cucur-bitaceae	Climber	Temperate	Chormi	1800–2200	Sep–Dec	Bark, seed	Paste, powder	Pneumonia, bronchitis, cardiac complaints, and as an anti-allergic.	—

NT - near threatened: EN - endangered: CREN - critically endangered: VU - Vulnerable toothache, *Thalictrum foliolosum* as a carminative, *Trichosanthes tricuspidata* for diabetes, *Paeonia emodi* for diabetes and blood dysentery, *Rheum australe* for internal injury, *Saussurea costus* for asthma, and *Saussurea gossipiphora* for cardiac complaints. This investigation elucidates several indigenous practices of medicinal plants in the adjoining area of NDBR and would be useful to naturalists, researchers, environmentalists, phytochemists, and practitioners of the indigenous system of medicine. It highlights new and noteworthy information on medicinal plants that can be utilized to uplift the economy of the inhabitants of Nanda Devi Biosphere Reserve.

It was also observed during the study that the natural resources are dwindling due to recent environmental changes in Garhwal Himalaya in particular and the Himalayas in general. Due to changes in the snow line (seasonal and permanent) and altitudinal shifting of tree lines, plant species are also shifting upwards to higher altitudes (Rana, Bisht, and Rana 2010). The nearest and easily accessible natural resources in the form of medicine are under threat. In this context, the age-old knowledge of the local and traditional system of health care is disappearing rapidly because the younger generation is less interested in the traditional and local primary health-care system. Not enough attention has been paid by legal professionals, environmentalists, non-governmental and governmental organizations to secure Intellectual Property Rights (IPR) and Natural Resource Property Right (NRPR) for the native people. The benefits accruing from the traditional knowledge systems should be shared with the communities or particular groups having this age-old tradition.

CONCLUSION

Today, although modern civilization is on a high pedestal in the field of medicine with its treatment of serious ailments, these facilities have not reached people who live far away from the towns; hence, such societies still mainly depend on traditional methods of treatment. It has been seen that, in some cases, modern science is still not able to treat some chronic ailments. The available resources and techniques in the traditional system of therapy for curing some of the chronic diseases may help improve the health-care system. It was observed that most of the *vaidyas* who possessed the traditional knowledge were more than 70 years old. If something is not done urgently to retrieve this precious knowledge, a day will come when this science will be lost forever with the passing of these people. There is also a need to establish home herbal gardens to make the required plant material available and to conserve the wild resources.

ACKNOWLEDGEMENTS

The authors are grateful to the inhabitants of the Nanda Devi Biosphere Reserve for providing information and hospitality. The CSR appreciates Dr Rakesh C Sundriyal, Director, Herbal Research and Development Institute, Mandal-Gopeshwar, for his critical evaluation of the manuscript and encouragement.

REFERENCES

Ansari, M. A., and Ghananand. 1985. Some medicinal plants of Pauri, Garhwal. *Himalayan Chemical Pharmceutical Bulletin* 2:42–44

Badoni, A. K. 1987–1988. Ethnobotany of hill tribes of Uttarkashi: plants used in rituals and psychomedicinal practices. *Journal of Himalayan Studies and Regional Development* 11,12:103–115

Badoni, A. K. 1989. Remarks on the high altitudinal medicinal plants of Garhwal Himalaya. *Journal of Himalayan Studies and Regional Development* 13,14:37–45

Badoni, A. K. 1990. An ethnobotanical study of Pinsari community: a preliminary survey. *Bulletin of Botanical Survey of India* 32:103–115

Badoni, A. K., and K. Badoni. 2001. Ethnobotanical heritage of Garhwal Himalaya. In *Nature, Culture and Society*, edited by O. P. Gusain and O. P. Kandari, pp.127–147. Garhwal: TransMedia

Balodi, B. S., and P. K. Hajra. 1995. *Plant Wealth of Nanda Devi Biosphere Reserve*. Howrah: Botanical Survey of India

Bhandari, B. S., R. D. Gaur, and D. S. Rawat. 2001. Vegetation wealth of Garhwal Himalaya. In *Nature, Culture and Society*, edited by O. P. Gusain and O. P. Kandari. Garhwal: TransMedia

Bhatt, K. C., and R. D. Gaur. 1992. A contribution to ethnobotany of Rajis in Pithoragarh district. *Acta BotanicaIndica* 20:76–83

Bhatt, K. C., M. K. Bisht, and R. D. Gaur. 1988. Folk medicine of Arakot valley in district Uttarkashi: an ethnobotanical study. In *Indigenous Medicinal Plants*, edited by P. Kaushik. New Delhi: Today and Tomorrows Printers and Publishers

Bisht, M. P. S., C. S. Rana, and V. Rana. 2010. An unusual composition of plant species towards zone of ablation (Tipra glacier), Garhwal Himalaya. *Current Science* 99:574–577

Burkill, I. H. 1965. *Chapters on the History of Botany in India*.Kolkata: Manager of Publication, Government of India

Chaudhari, U., J. R. Sharma, D. K. Singh, and B. P. Uniyal. 2007. *Flowering Plants of Uttarakhand (A Checklist)*. Dehradun: Bishan Singh Mahendra Pal Singh

Dangwal, L. R., C. S. Rana, and A. Sharma. 2010. Ethno-medicinal plants of the Garhwal Himalaya used to cure various diseases: a case study. *Journal of American Science* 3:12–16

Dangwal, L. R., C. S. Rana, and J. K.Tiwari. 2010. Ethnopharmacology, indigenous collection of some important medicinal plants of Nity Valley in Nanda Devi National Park, Uttarakhand. *Nature and Science* 8(11):9–14

Dangwal, L. R., N. Kumar, C. S. Rana, A. Sharma, and U. Sharma. 2010. Ethno-medico botany of some aquatic Angiospermae from north-west Himalaya. *Researcher* 2(4):49–54

Dangwal, L. R., R. Kuniyal, C. S. Rana, J. K. Tiwari, and P. Tiwari. 2010. Ethnomedicinal plants of Nanda Devi Biosphere Reserve, Uttarakhand. *Reports and Opinion* 2(2):58–61

Deva, S., and H. B. Naithani. 1987. *The Orchid Flora of North West Himalaya*. New Delhi: Print & Media Associates

Dhasmana, H., B. Lal, and R. K. Nigam. 1989. A contribution to the medicinal plant-lore of Garhwal region. *Himalayan Research and Development* 8(182):24–30

Gaur, R. D. 1999. *Flora of the District Garhwal: Northwest Himalaya* (*with Ethnobotanical Notes*). Garhwal: TransMedia

Gaur, R. D., and J. K. Tiwari. 1987. Some little known medicinal plants of Garhwal Himalaya: an ethnobotanical study. In *Medicinal and Poisonous Plants of the Tropics*, edited by A. G. M. Leeuwenberg, pp.139–142. Pudoc: Wageningen

Gaur, R. D., J. K. Semwal, and J. K. Tiwari. 1984. A survey of high altitude medicinal plants of Garhwal Himalaya. *Bulletin of Medicinal Ethnobotany Registration* 4(3–4):102–116

Gaur, R. D., K. Negi, K. C. Pant, and J. K. Tiwari. 1993. Notes on ethnobotany of five districts of Garhwal Himalaya. *Ethnobotany* 5:73–81

Gaur, R. D., K. S. Negi, and J. K. Tiwari. 1985. Ethnobotanical importance of some trees in Garhwal Himalayas. *Journal of Indian Botanical Society* 61:32

Hajra, P. K., and B. Balodi. 1995. *Plant Wealth of Nanda Devi Biosphere Reserve*. Kolkata: Botanical Survey of India

Jain, S. K., and R. R. Rao. 1997. *A Handbook of Field and Herbarium Methods*. New Delhi: Todays and Tomorrow's Printers and Publishers

Joshi, H. C., and S. S. Samant. 2003. Floristic diversity, community pattern and changes of vegetation in Nanda Devi National Park. In *Biodiversity Monitoring Expedition*, pp. 39–44. Uttaranchal: Forest Department

Joshi, P., and P. C. Pande. 1995. Ethnobotany of the Rajis of Kumaun Himalaya: a preliminary study. In *Rishi-II. Uttaranchal Himalaya*, edited by M. P. Joshi, pp. 241–263. Almora: Shree Almora Book Depot

Joshi, P., and P. C. Pande. 1997. Ethnobotany of Bhotia tribe of Kumaun Himalaya. In *Studies on Kumaun Himalaya*, edited by N. C. Dhoundiyal and S. K. Sharma, pp. 63–89. New Delhi: Indus Publishing

Kala, C. P. 2000. Status and conservation of rare and endangered medicinal plants of Indian Trans-Himalaya. *Biological Conservation* 93:371–379

Kala, C. P. 2002. *Medicinal Plants of Indian Trans-Himalaya*. Dehradun: Bishen Singh Mahendra Pal Singh

Kala, C. P. 2005. Indigenous uses, population density, and conservation of threatened medicinal plants in protected areas of Indian Himalayas. *Conservation Biology* 19:368–378

Lal, B., A. Saklani, and H. Singh. 1990. Ethnobotanical observations on some gymnosperms of Garhwal Himalaya. *Economic Botany* 44:349–354

Maheshwari, J. K., S. K. Mumgain, and R. R. Rao. 1990. Some medicinal plants of Pauri Garhwal. *Journal of Economic and Taxonomic Botany* 14:633–640

Maheshwari, J. K., and J. P. Singh. 1984. Contribution to the ethnobotany of Bhoxa tribe of Bijnor and Pauri Garhwal district, UP. *Journal of Economic and Taxonomic Botany* 5:253–259

Maikhuri, R. K., S. Nautiyal, K. S. Rao, and K. G. Saxena. 1997. Medicinal plants cultivation and biosphere reserve management: a case study from the Nanda Devi Biosphere Reserve, West Himalaya. *Current Science* 74:157–163

Maikhuri, R. K., S. Nautiyal, K. S. Rao, and K. G. Saxena. 1998. Role of medicinal plants in the traditional health care system: a case study from Nanda Devi Biosphere Reserve. *Current Science* 75:152–157

Mumgain, S. K., and R. R. Rao. 1990. Some medicinal plants of Pauri Garhwal. *Journal of Economic and Taxonomic Botany* 14:633–640

Murti, S. K. 2001. *Flora of Cold Deserts of Western Himalaya*. Kolkata: Botanical Survey of India

Naithani, B. D. 1984. *Flora of Chamoli*, vol. 1. Kolkata: Botanical Survey of India

Naithani, B. D. 1985. *Flora of Chamoli*, vol. 2. Kolkata: Botanical Survey of India

Nautiyal, S. 1981. Some medicinal plants of Garhwal hills: a traditional use. *Journal of Scientific Research on Plants and Medicines* 2:12–18

Negi, K. S., and R. D. Gaur. 1994. Principal wild food plants of Western Himalaya, UP, India. In *Higher Plants of Indian Subcontinent*, edited by B. K. Gupta, vol. III. pp. 1–78. Dehradun: Bishen Singh Mahendra Pal Singh

Negi, K. S., and K. C. Pant. 1992. Less known wild species of *Allium* Linn (Amaryllidaceae) from mountainous region of India. *Economic Botany* 46:112–116

Pangtey, Y. P. S., G. S. Rawat, and S. S. Samant. 1989. Ethnobotanical notes on Bhotia tribes of Kumaun Himalaya. *Indian Journal of Forestry* 12:191–196

Rana, C. S. 2007. Ethnobotanical studies on the medicinal plants of Nanda Devi Biosphere Reserve, Uttaranchal. [PhD thesis submitted to Garhwal University, Srinagar, Garhwal]

Rau, M. A. 1975. *High Altitude Flowering Plants of West Himalaya*. Kolkata: Botanical Survey of India

Tiwari, J. K. 1986. Ethnobotanical study of the medicinal plants of Garhwal. [PhD thesis submitted to Garhwal University, Srinagar, Garhwal]

Uniyal, B. P., J. R. Sharma, U. Chaudhari, and D. K. Singh. 2007. *Flowering Plants of Uttarakhand (A Checklist)*. Dehradun: Bishan Singh Mahendra Pal Singh

CHAPTER 15

Indigenous Knowledge, Use of NTFPs, and Biodiversity Conservation in Uttarakhand

Chandra Prakash Kala

INTRODUCTION

Non-timber forest products (NTFPs) have gained tremendous importance at present, and is evident from the fact that the Conference of Parties (COP-6) to the Convention on Biodiversity (CBD) that took place in The Hague, Netherlands, during 7–26 April 2002 had major discussions on medicinal plants, indigenous knowledge, and benefit sharing. Since the Indian subcontinent is well known for its diversity of forest products, including the age-old valuable traditions associated with them, the need of the hour is to establish these traditional values in both the national and the international perspectives realizing the ongoing developmental trends in traditional values across the globe (Kala 2010a).

The NTFPs have become mainly an alternate source of income for the the poor, and therefore have the potential to benefit and improve the living standards of the poor people. A great deal of traditional knowledge on NTFP is still intact with the indigenous people, especially those living in the mountainous areas. The business opportunities are restricted due to poor accessibility of the terrain and comparatively fewer developmental activities, hence enabling the continuity of indigenous knowledge as a way of life in these areas (Kala 2007). The various indigenous communities living here in the Himalayan region are repositories of rich traditional knowledge, especially on the uses of plant species for various purposes. It, thus, indicates their general awareness and intimate dependence on the surrounding plant species (Gaur 1999; Kala 2004). Such traditional knowledge is especially vital in today's context for various reasons, including population explosion and growing demand for food, clothing, medicine, and shelter. At the same time, many indigenous practices are getting recognition as a valuable tool for environmental conservation and management.

Considering these facts and realizing the continuous erosion of the old traditional customs, skills, and beliefs, it is important to document this valuable knowledge before it gets completely lost (Kala 2000). This documentation of the indigenous knowledge of the Bhotiya community on the uses of forest products in Uttarakhand is an attempt to save such knowledge for the future and make it more relevant through value addition and adaptation.

STUDY AREA

This study was conducted in Uttarakhand, which is located in the northern region of India and covers an area of 53 485 km^2 (Figure 1). The population of the state is 8479562, of which 78% resides in rural areas. The state is known for its rich biodiversity, culture, tradition, and mythology. The large altitudinal differences from the Himalayan foothills to the high peaks, and, thus, the climatic and topographical variations have supported and maintained the various forest types (tropical, temperate, and sub-alpine) in the state.

METHODOLOGY

Field surveys were undertaken across the various districts of Uttarakhand. A semi-structured questionnaire survey was conducted among selected knowledgeable people, including the traditional *vaidyas*, to document their knowledge on the preparation of

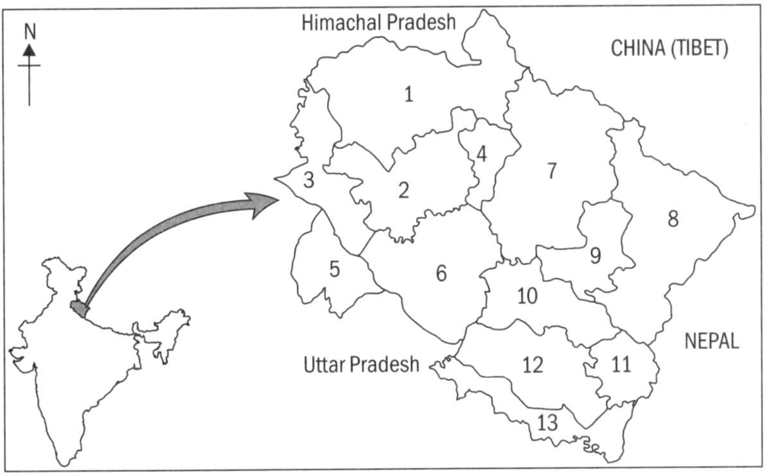

Figure 1 *Location of study area (Uttarakhand, India)*

Note *Numerical values indicate the districts in the state. 1 – Uttarkashi, 2 – Tehri, 3 – Dehradun, 4 – Rudraprayag, 5 – Haridwar, 6 – Pauri, 7 – Chamoli, 8 – Pithoragarh, 9 – Bageshwar, 10 – Almora, 11 – Champawat, 12 – Nainital, 13 – Udham Singh Nagar*

indigenous medicinal formulations, beverages, and food including the use of plants for such purposes. Information was also gathered on the use of plant parts. The data was cross-checked by interviewing a number of elderly and knowledgeable respondents on the uses of specific plant species and the preparation of particular formulations, food, beverages, and natural dyes. Participant observation was also used to understand the indigenous methods and techniques adopted by the local people for preparing food, beverages, and medicine. Interactive workshops were organized in different districts of Uttarakhand and various groups of indigenous people including *vaidyas* were invited to generate and help in documenting indigenous knowledge on preparing food, beverages, and medicine.

RESULTS AND DISCUSSION

Indigenous Knowledge and Uses of Medicinal Plants

A total of 243 formulations, prepared by different traditional *vaidyas*, have been documented. These formulations, made of 156 medicinal plants, are used to cure 73 different ailments. Most of the formulations are documented for curing coughs and colds, followed by headaches, indigestion, dysentery, and toothaches. There is a sharp decline in the number of recognized *vaidyas*, but there are a number of women and men in the villages who know the healing properties of different medicinal plants. The loss of knowledge on preparing medicine is due to the decline in the number of *vaidyas* coming forward to adopt this traditional healing practice professionally. At present, the knowledge is scattered among the local inhabitants (Kala 2005a, 2006).

The study indicates that the fundamental principles that the *vaidyas* use to prepare herbal medical formulations are often combined with philosophy, religion, and various other ingredients. On the basis of nature and origin, the drugs are classified into three groups: (1) vegetables, (2) animals, and (3) metals/minerals. Drugs of vegetable origin are further divided into four groups based on the plant parts used, which are (1) fruits, (2) flowers, (3) creepers, and (4) annuals. The drug is developed and given to the patient in the form of juice, powder, decoction, paste, jam, or pills. The *vaidyas* system of medicine pursues the holistic approach and aims not only to cure the affected organ alone but also to find out the origin and cause of the disease to eradicate it from its root (Kala 2006).

The results of the study indicate that most of the common species of plants that grow in gardens and adjacent to the village areas in cultivated or forested lands are used in the preparation of herbal medical formulations by the *vaidyas* in most cases. *Ocimum sanctum*, *Piper nigrum*, *Curcuma domestica*, *Brassica compestris*, and *Raphanus sativus* are frequently used to prepare the medicine. The plants are collected for medicinal use when they attain maturity, judged by their height, patterns of branching, colour, and other morphological characteristics including the right fragrance and potency. These judgments are mainly based on experience and knowledge of the collectors. There are also some guidelines for collecting different parts of the medicinal plant species. The branches are collected when they are fully grown during spring and the rains. Young leaves are collected mainly between the flowering stage and the ripening of fruits. If the roots of the trees are required, then only the bark of hard and woody roots is collected. Roots and rhizomes of annuals are collected in summer and winter after the leaves fall or when the new leaves are just emerging. Milk, sap, gum, resin, latex, and other liquid exudes from plants are collected in autumn depending on their availability. Similarly, the flowers and fruits are collected according to the seasonality and availability (Kala 2010b, 2010c).

Other plant species are mixed into the formulation sometimes in order to counter possible side effects from the actions of the primary plants. However, there are many medicinal herbs such as Katuki (*Picrorhiza kurooa*), Hardi (*Terminalia chebula*), Haida (*Terminalia bellarica*), Brahmi (*Centella asiatica*), Pudina (*Mentha longifolia*), Haldi (*C. domestica*), and Ashwagandha (*Withania somnifera*) that are given alone to cure ailments. The *vaidyas* use the whole herb instead of extracting the active ingredient from the plant. The notion of using the whole herb rather than the isolated ingredient may also contribute to a balanced formula less likely to have side effects (Kala 2010b).

Apart from this, a total of 43 species are used as a tonic. Many of these species are currently in different threat categories as per Conservation Assessment and Management Plan (CAMP) workshops. For instance, six species are placed in critically endangered, three in endangered, and two in vulnerable categories. For vitalizing hair, of eight species: *Arnebia benthamii*, *Arnebia euchroma*, *Cuscuta reflexa*, and *Ficus benghalensis* are the most important. For general health tonics, of the total 37 species, *Dactylorhiza hatagirea*, *Polygonatum verticillatum*, *Nardostachys grandiflora*, *Fritillaria roylei*, *Eulophia dabia*, and *Aegle marmelos* are the most frequently used (Kala 2003).

Among the brain tonics, *C. asiatica* is one of the important species that has been mentioned in old literature by sages by the name of Brahmi. The local people of Uttarakhand considered this species as a cold medicine, which nourishes the brain. This species is also being used to prepare cold drinks. The rhizome of *P. verticillatum* has a reputation for being an excellent tonic for general health care. It has been mentioned as a nervine tonic by Jain (1991). *N. grandiflora*, a critically endangered species, is also used as a health tonic. It grows only in the high altitudes of the Himalaya. The rhizome of this species is used as a substitute for valerian and the extract finds use in 26 ayurvedic preparations (Prakash 1999). It is also used in curing various heart diseases and is also good as a heart tonic. This is an important medicinal plant but, due to overexploitation, its population is shrinking in the wild.

The terrestrial orchid *D. hatagirea* (Hathajari) is the best-known tonic for health care as its tuber is used in preparing Salep, a very popular tonic in the UK. Germany is one of the major importers of Salep (Kasparek and Grimm 1999). In Turkey, Salep is derived from at least 30 species of orchids and supplied mostly to Germany. In Uttarakhand, people collect *D. hatagirea* from alpine and sub-alpine meadows and use the entire tuber as Salep.

Apart from the common formulations used by the *vaidyas*, a special herbal drug is used by some of them in the treatment of leucorrhea. About 24 species of medicinal plants are mixed in a definite proportion to prepare this herbal formulation. *Mucuna pruriens*, *Plantago depressa*, *Chlorophytum tuberosum*, *Tinospora cordifolia*, *A. marmelos*, *Holarrhena antidysenterica*, and *Curculigo orchioides* are the major species used, in high proportions, to prepare this. Besides, some endangered medicinal plant species, such as *Nardostachys jatamansi*, *D. hatagirea*, and *Crocus sativus* are also used for enhancing the efficacy of this herbal drug (Kala 2010b).

The list of plants used in the treatment of different ailments in Uttarakhand is diverse. Even within the minor altitudinal variations, one finds diverse uses of the same plant species. Although some plant species are used to cure only a few diseases, they are known for their quick and effective healings. There is a lot of pressure on the natural medicinal plant population due to the high demand and exploitation (Kala 2005b). Among the various plant parts used in therapy, the parts that grow underground, such as the rhizome, tuber, runner, and root, are used extensively. This has increased the pressure on the survival of such wild populations. Due to the restricted distribution of these medicinal plants, both illegal and legal collections are being grown in these alpine meadows.

There is no doubt that some medicinal plants are critically endangered due to their small population, narrow range of distribution, habitat specificity, destructive mode of harvesting, high-use values, over-collection for commercial purpose, habitat alteration, and present developmental activities. Recent studies provide some information on frequency of use, but gathering and transmitting the right information on the extent of trade is difficult due to lack of transparency and accountability of the groups involved in the collection and supply of medicinal plants for trade. A two-pronged strategy needs to be evolved—one that takes care of the in-situ efforts, such as the implementation of a medicinal plant conservation area (Kala 2000) and another that promotes the cultivation of medicinal plants through both conventional and biotechnological tools. Both parts of the strategy are not only important with regard to conservation and sustainable use of the resources but will also, more importantly, preserve the indigenous knowledge of various ethnic groups about herbal remedies (Silori and Badola 2000; Kala 2002).

Indigenous Knowledge on Fermented Food and Beverages

The traditional starter culture used in the preparation of fermented food and beverages is called *balam* in Kumaon and *balma* in Garhwal, and is not prepared by all villagers. This starter culture is made by mixing wheat with a number of herbs and spices, such as cinnamon *(Chinnamomum zeylanicum)*, black cardamom *(Amomum subulatum)*, black pepper *(Piper longum)*, leaves of mirchi-ghash (wild chillies), and seeds of sacred fig (*Ficus religiosa*). The addition of old *balam* powder is a necessity, without which the production of fresh *balam* is not possible (Roy, Kala, Farooquee, *et al.* 2004; Kala 2004).

Jaan is a traditional drink of the Bhotiyas, made commonly from rice. However, it can also be made from a good number of substrates of both cereals and fruits. Some cereals from which *jaan* can be made are rice (*Oryza sativa*), wheat (*Triticum aestivum*), barley (*Hordeum vulgare*), foxtail millet (*Setaria italica*), common millet (*Panicum miliaceum*), *oowa* (*Hordeum himalayens)*, and *chuwa* (*Amaranthus paniculatus*). Among fruits, apple is most desired when making a delicious *jaan*. But *jaan* prepared from foxtail millet is considered the best in quality. The quality of *jaan* is best judged by its taste (sweetness), smell, and strength (Roy, Kala, Farooquee, *et al.* 2004).

Daru is a distilled liquor that contains ethyl alcohol at a much higher concentration than other alcoholic beverages. Rice and jaggery are the common substrates for preparing *daru*. Apart from rice, cereals like *kauni*, *chuwa*, *oowa*, and wheat are also used. However, unlike *jaan*, the taste (quality) of *daru* does not vary with the type of substrate used. Therefore, the choice of substrate does not matter in its preparation. The most commonly available and cost-effective materials used to make *daru* are rice and jaggery.

The traditional semi-fermented food used by the Bhotiyas is called *sez*. It is made from rice, and is mostly used as a snack. Earlier, it was a delicacy and made only during certain festivals. In most cases, *sez* is extracted during the preparation of rice *jaan* (*chaul-ki-jann*). In the case of *daru* preparation, the intermediate stage that yields *sez* is unstable. The quantity of *balam* powder added determines the rate of fermentation. When preparing *daru*, the maximum quantity of *balam* is used. Thus, whenever the extraction of *sez* is required, the fermentation process has to be slowed down. To that effect, a small quantity of *balam* powder is mixed with the substrate. During slow fermentation, it becomes easy to intervene in the process and remove the *sez*. Once the *sez* has been removed, fresh *balam* powder is added to hasten the process of fermentation for *jaan* or *daru*.

The Bhotiyas have adapted to the emerging and changing climatic conditions in the region and can survive on the surrounding natural resources. However, there is a shift in the lifestyle due to the intervention of external forces, which have been damaging the traditional and self-sustaining systems in the name of development, through market-driven products. The indigenous knowledge of making fermented food and beverages is a traditional practice developed and practiced for generations. However, over a period of time, the expansion of the road network, market forces, availability of ready yeast, and modern liquor have influenced the quality and quantity of indigenous fermented food and beverages. The age-old fermentation techniques should be encouraged, as they lead to the development of nutritious food items, which help in coping with the inhospitable climatic conditions of high-altitude areas.

Indigenous Knowledge on Wool Dyeing

Eight plant species: *Rheum australe* (Dolu), *Rumex nepalensis* (Khukhuyinya), Bajar Bhang, *Berberis* sp. (Kingod), *Juglans regia* (Akhrot), *Corylus jacquamontii* (Kapasi), *Geranium nepalense*, and a few lichens are recorded as used by the Bhotiya ethnic groups to prepare various natural dyes. The root and the fruit cover are the plant parts most used for this. The Bhotiya tribes use the roots of five plant species and fruit covers of two. The colours extracted from these species are yellow, red, camel, coffee brown, and mud brown. By mixing these base colour dyes in different combinations, a number of intermediate colours and shades are produced. For example, *J. regia* and *R. australe* are mixed together to produce red (Kala 2002).

Of these eight plant species, *R. australe* and *Rheum moorcrotianum* are vulnerable while *Berberis aristata* is endangered as per the new IUCN criteria. Both, *R. australe* and *R. moorcrotianum* are herbaceous species and generally grow above the timberline in the Himalayas, whereas *B. aristata* is a shrub that grows in the middle and lesser Himalayan areas. As these three species also possess high medicinal properties, there is a great demand for them from pharmaceutical companies and herbal healers. There is a ban on collecting the species from both protected and unprotected areas. Therefore, the Bhotiya community faces a problem in obtaining the raw material to continue the indigenous practice of wool dying (Kala 2002).

CONCLUSION

The traditional values attached with various NTFPs have gained tremendous importance in the present era. This is evident from the fact that the agendas of the Conference of Parties to the Convention on Biodiversity include strengthening indigenous knowledge systems, the medicinal plant sector, and benefit sharing issues. Since the Himalayan region of India is famous for its variety of ethnobotanical species as well as the age-old, valuable indigenous knowledge associated with them, the need of the hour is to establish these traditional values in both national and international perspectives, keeping in mind the present developmental trends. The indigenous communities of Uttarakhand have evolved and established various mechanisms for utilizing the available natural resources of plants for medicine, fermented food and beverages, and wool dyeing. The traditional method of making fermented food and beverages is designed by these people in such a way that it is utilized for the available crops. These indigenous communities are replete with such examples of prudence in the management and utilization of available natural resources.

REFERENCES

Gaur, R. D. 1999. *Flora of the District Garhwal Northwest Himalaya (with Ethnobotanical Notes)*. Srinagar, Garhwal: TransMedia

Jain, S. K. 1991. *Dictionary of Indian Folk Medicine and Ethnobotany*.New Delhi: Deep Publications

Kala, C. P. 2000. Status and conservation of rare and endangered medicinal plants in the Indian trans-Himalaya. *Biological Conservation* 93(3):371–379

Kala, C. P. 2002. Indigenous knowledge of Bhotiya tribal community on wool dyeing and its present status in the Garhwal Himalaya, India. *Current Science* 83(7):814–817

Kala, C. P. 2003. Indigenous uses of plants as health tonic in Uttaranchal Himalaya, India. *Annals of Forestry* 11(2):249–54

Kala, C. P. 2004. *Studies on the Indigenous Knowledge, Practices and Traditional Uses of Forest Products by Human Societies in Uttaranchal State of India*. Almora: GB Pant Institute of Himalayan Environment and Development

Kala, C. P. 2005a. Current status of medicinal plants used by traditional *vaidyas* in Uttaranchal state of India. *Ethnobotany Research and Applications* 3:267–278

Kala, C. P. 2005b. Indigenous uses, population density, and conservation of threatened medicinal plants in protected areas of the Indian Himalayas. *Conservation Biology* 19(2):368–378

Kala, C. P. 2006. Preserving Ayurvedic herbal formulations by *vaidyas*: the traditional healers of the Uttaranchal Himalaya region in India. *HerbalGram* 70:42–50

Kala, C. P. 2007. Local preferences of ethnobotanical species in the Indian Himalaya: implications for environmental conservation. *Current Science* 93(12):1828–1834

Kala, C. P. 2010a. Ethnobotanical and ecological approaches for conservation of medicinal and aromatic plants. *Acta Horticulturae* 860:19–26

Kala, C. P. 2010b. *Medicinal Plants of Uttarakhand: Diversity Livelihood and Conservation.* Delhi: Biotech Books

Kala, C. P. 2010c. Assessment of availability and patterns in collection of Timroo (*Zanthoxylum armatum* DC.): a case study of Uttarakhand Himalaya. *Medicinal Plants* 2(2):91–96

Kasparek, M., and U. Grimm. 1999. European trade in Turkish Salep with special reference to Germany. *Economic Botany* 53:396–406

Prakash, V. 1999. *Indian Valerianaceae: A Monograph on Medicinally Important Family.* Jodhpur: Scientific Publisher

Roy, B., C. P. Kala, N. A. Farooquee, and B. S. Majila. 2004. Indigenous fermented food and beverages: a potential for economic development of the high altitude societies in Uttaranchal. *Journal of Human Ecology* 15(1):45–49

Silori, C. S., and R. Badola. 2000. Medicinal plant cultivation and sustainable development: a case study in the buffer zone of the Nanda Devi Biosphere Reserve, Western Himalaya, India. *Mountain Research and Development* 20(3):272–279

Section III

Climate Change and Biodiversity Conservation

CHAPTER 16

Climate Change and Challenges of Biodiversity Conservation

Chandra Prakash Kala

INTRODUCTION

Human beings are dependent on biological resources for food, cloth, health, and enjoyment of life; therefore, depletion of these valuable resources is the cause of great concern. Unfortunately, there has been an unprecedented loss of biodiversity worldwide, and it is mainly attributed to anthropogenic activities due to population explosion. To decline the rate of biodiversity loss, protected areas have been created across the globe. However, the impact of climate change on biodiversity cannot be diminished by creating fences and demarcating boundaries for biosphere reserves, national parks, and sanctuaries (Graves and Reavey 1996; Kala 2005). In the recent decades of conservation biology research, some consensus has emerged on 'what' to conserve and 'how' to conserve (Redford, Coppolillo, Sanderson, *et al.* 2003). However, the approach for 'how' to conserve is still an issue open for debate (Bawa, Seidler, and Raven 2004). The existing knowledge on the effects of global climate change on biodiversity is meagre and also not accepted unanimously by different disciplines. This chapter aims to synthesize and analyse diverse opinions and observations made by different schools of thought on the various impacts of climate change on biodiversity, and the suggested mitigation strategies.

EFFECTS OF CLIMATE CHANGE ON BIODIVERSITY

Biodiversity is known to be influenced by the change in climate across the globe. Despite this fact, the opinion on climate change and its effects on biodiversity is not accepted unanimously; there are arguments and counter arguments. One school of thought argues that the effects of climate change on biodiversity are lower than the effects of land-use changes and intrinsic properties of biodiversity (Lomborg 2001). This observation seems to ignore the potential complexity of natural systems and biotic changes. Change and extinction of even one species may have profound effect on the other species, as is evident from the extinction of the dodo, a flightless bird. After the bird became extinct, the seeds of the Mauritian calvaria tree stopped sprouting and it appeared that the tree might also become extinct. The short-term local changes may become important in the longer term if the causal factor continues. The small but constant forces are important because they can alter species interactions and destabilize communities (Parmesan and Yohe 2003).

The evidence indicates that changes in plant phenology, range and distribution of species, composition and interactions within communities, and even changes in structure and dynamics of ecosystems occur due to climate change (Walter, Post, Convey, *et al.* 2002). Changes in the timing of seasonal activities of species are reported to occur due to climate change. Some of such evidences are the advancement of the flowering season and breeding of a few species, including butterflies and amphibians, and early migration by migratory birds (Graves and Reavey 1996; Walter, Post, and Convey, *et al.* 2002; Mawdsley, O'Malley, and Ojima 2009). Species may not obtain suitable habitats and food for survival due to local changes in climatic conditions and the different patterns in associated species that may not be synchronous, resulting in the alteration of the ecosystem's properties.

The increasing temperature and subsequent changes in the resource availability may shift species poleward in latitude and upwards in elevation. Studies show a typical pattern in species distribution, which include northward range shifts during the two 20th-century warming periods (1930–45 and 1975–99), and southward shifts during the intervening cooling period (1950–70) (Parmesan and Yohe 2003; Parmesan 2006). Tree species, such as pine and maple, which depend on the wind to disperse their seeds, might become extinct due to high CO_2 concentration and reduction of surface wind speed (Nathan, Horvitz, He, *et al.* 2011). In the Himalayan region of India, oak forests are being invaded by pine forests due to the increase in temperature. The associated species of pine forest may cross frontiers and become new elements of the oak forests. There are phenological changes in terms of early flowering in rhododendrons, one of the major companion species of oak forests in the Himalayas. The treeline zone in the Himalaya is known for its high sensitivity to climate change, and increasing temperature may have a profound influence on the floral and faunal diversity of this important ecosystem. Despite many studies, projecting and predicting the right direction and the impact of climate change on species distribution remain challenging tasks to be tackled.

It is an established fact that increasing CO_2 concentration increases plant growth (Graves and Reavey 1996). However, photosynthesis is a fine-tuned process and depends on the availability of nutrients in the soil and the temperature of the environment. For example, since the high-altitude areas of the Himalaya are generally poor in nutrients, climate change may have a negative effect on the rate of photosynthesis in such areas. From this region, many of the plant species are used to prepare medicines. Any change in plant properties, such as a change in secondary metabolites due to climate change, may severely impact the industries dealing with this plant produce (Gairola, Shariff, Bhatt, *et al.* 2010). Change in vegetation distribution may not guarantee a shift in the companion plant and animal species. The less mobile fauna are especially assumed to be more affected due to change in climate.

The effects of climate change on biodiversity can be noticed at all three levels of diversity—genetic, species, and ecosystem. The changing climatic conditions may alter the deoxyribonucleic acid (DNA) structure and chromosome arrangements of species

(Rezende, Balanyà, Rodríguez-Trelles, *et al.* 2010). Depending on the vigour of newly evolved genetic variations and their capacity to adapt to prevailing climatic conditions, the species may evolve and survive or become extinct. The climate change may affect species diversity in several ways. Species may tend to move to areas with more suitable climate, within a certain tolerance level, leading to a change in their distribution. Species with low tolerance may adapt or modify themselves to new climatic conditions, creating a new range of species. Species seldom have any option but to become extinct if they cannot tolerate or adapt to the new environmental system (Graves and Reavey 1996). Species acclimatized to specific habitats or microhabitats may become extinct due to specialized habitat requirements. Interactions between species of plants and animals and amongst themselves may also change owing to the change in species characteristics and distribution. The life cycle of species will change, as temperature affects their birth, growth, and reproductive periods. The community composition in a given area as well as the ecosystem may alter due to the change in frequency and density in population. Variations in resultant ecosystem properties may affect the whole range of species with maximum stress on endemic species, which may lead to local extinction of the species.

The adaptive capacity and the sensitivity of ecosystems and biodiversity to climate change may differ across the biome; therefore, the effect of climate change would vary accordingly. Although multiple factors influence the ecosystem and biodiversity, climate change is considered a prime factor that changes the biodiversity in an Arctic ecosystem (Sala, Chapin, Armesto, *et al.* 2000). The biodiversity of cold climatic areas, such as alpine and sub-alpine regions of the Himalayas, is, therefore, a matter of great concern in the present era of global climate change. Apart from forests and wildlife, agro-biodiversity is getting influenced by the climate change. Although the modern agriculture system has increased food production, it has damaged the natural environment and biodiversity with the increased use of chemical fertilizers. The unsustainable dependence on natural resources, particularly on ground water, has caused water scarcity and, in some cases, made land unsuitable for cultivation due to salinity ingress (Kala 2010a).

Climate Change, Protected Areas, and Biodiversity Conservation

A key issue worldwide in the 21st century is understanding the effect of climate change on our planet. Studies have been carried out to determine the effects of climate change on both marine (Hays, Richardson, and Robinson 2005) and terrestrial ecosystems (Nemani, Keeling, Hashimoto, *et al.* 2003). There are predictions of climate change on natural ecosystems and biodiversity, but there is poor consensus on the effects among the various disciplines, due to contentious research results. The large gap in our understanding and knowledge on biodiversity and climate change has given way to shape different opinions on spreading greenery on Earth over the past few decades. We are not sure about the outcomes of biodiversity conservation within tropical ecosystems, where much of the world's biodiversity resides.

A study that evaluated tropical forests concludes that, in warm years, trees are growing more slowly and many of them are dying (Kaiser 2003), as a result of which the increase in CO_2 concentration and temperature could lead to low net primary production. At the same time, another study states that, from 1982 to 1999, global changes in climate have eased several critical climatic constraints on plant growth, in such a way that net primary production increased 6% globally, and the ecosystems in all the tropical regions accounted for 80% increase in net primary production. This is reported to be due to increase in air temperature, cloudiness, changing monsoon dynamics, increasing incidence of solar radiation, and other climatic constraints (Nemani, Keeling, Hashimoto, *et al.* 2003).

Since there is a scarcity of multidisciplinary approaches to assess the natural and human-induced effects on biodiversity, the satellite-based estimates of net primary production by Nemani, Keeling, Hashimoto, *et al.* (2003) overlooked the effects of biodiversity conservation and protected area networks over the past three decades. At present, the global network of protected areas covers 11.5% of the Earth's land surface, which surpasses the 10% target proposed a decade earlier (Rodrigues, Andelman, Bakarr, *et al.* 2004). This large coverage of protected areas would, naturally, have been beneficial in safeguarding the various types of natural forests that were overexploited in the past, and now lead to an increase in net primary production.

Similarly, planting of various tree species over the past three decades in the world's terrestrial ecosystem would have increased the net primary production. Most of the protected areas have been established after the 1970s. As the study of Nemani, Keeling, Hashimoto, *et al.* (2003) was carried out after about a decade of introducing protected area networks, it also logically reflects the effects of protected area networks and other conservation policies such as plantations on productivity. The authors conclude their study by saying that assuming that climatic constraints or increase in CO_2 are responsible for increase in net primary production across global terrestrial ecosystems. However, without assessing other causal factors, including the effects of several years of forest protection, such conclusion seems premature (Kala 2005).

Besides the effects of climate change on biodiversity, the protected area policy is being questioned. For instance, during the last 20 years, the buffer zones of nearly 70% protected areas in tropical regions have reported a decline in their extent, while another 25% have reported a decline due to altered administrative boundaries (Defries, Hansen, Newton, *et al.* 2005). Before the creation of protected area networks, policymakers perceived that practices of natural resource utilization, such as livestock grazing, were largely responsible for loss of biodiversity. Hence, they established a law prohibiting livestock grazing in national parks. Contrary to this fact, there are ample studies stating that moderate grazing and mechanical mowing increase plant diversity probably by shifting the competitive balance among plant species, decimating plant populations, and weakening dominant species, thereby allowing others to thrive (Marquis 2004; Moore 2005). These studies suggest that the future ability of protected areas to maintain species richness depends on integrating reserve management within the land-use dynamics of their larger regional settings. Hitherto, the conservation targets and approaches are defined on large spatial scales hoping that they may apply broadly. As a result, a succession of generalized, monolithic conservation models have been seen replacing one another or competing with one another for attention (Bawa, Seidler, and Raven 2004).

It has been established that biodiversity enhances an ecosystem's reliability (Naeem and Li 1997), and declining biodiversity can alter the performance of ecosystems (Naeem, Thompson, Lawler, *et al.* 1994). However, the major gap in our understanding is that we have identified only a fraction of the earth's biological diversity and have just a rudimentary understanding of how biological, geographical, and geochemical processes interact to contribute to human well being (Balmford, Bennun, Brink, *et al.* 2005). This undeveloped knowledge and understanding of a small fragment of biodiversity have provided endless opportunities to speculate on many unrealistic tales (Storch, Marguet, and Gaston 2005).

CLIMATE CHANGE MITIGATION STRATEGIES AND CDM PROJECTS

To maintain ecological integrity, species that are at the risk of extinction need immediate conservation efforts, both ex situ and in situ. There are arguments on the suitability of species-oriented conservation planning and an ecosystem- or biome-based conservation approach. Besides, the effects of climate change are beyond the boundaries of protected areas such as national parks, sanctuaries, and biosphere reserves. Apart from habitats, the movement corridors of wildlife species need to be protected. Despite several reports and studies, it is a well-known fact that the data on the effects of climate change, especially on biodiversity, are limited. However, given the urgency, it is preferable not to wait for a solid data set to validate and calibrate models and concepts adequately for the protection of nature and biodiversity.

In pursuit of the major objective of the United Nations Framework Convention on Climate Change (UNFCCC), Article 2 of the Kyoto Protocol specifically deals with the protection and enhancement of sinks and reservoirs of greenhouse gases (GHG), which ultimately encourage sustainable forest management practices including afforestation and reforestation. To stabilize GHG emissions and, hence, the effects of climate change on biodiversity, the state of knowledge on biodiversity needs to be evaluated. This will help to develop effective policies, regulations, and laws to mitigate the effects of climate change.

The essence of Clean Development Mechanism (CDM), which is an arrangement under Article 12 of the Kyoto Protocol of the UNFCCC, is to allow developed country parties to jointly undertake emission reduction project activities to meet their GHG reduction commitment. Launching CDM projects may generate awareness for environmental conservation and its activities, in partnership with foreign investors, provide an opportunity for local partners to learn new skills. The CDM projects may contribute to the sustainable development of countries through transfer of technology and knowledge, improving social and human capital by creation of employment, raising living standards, and improving environmental assets by preserving biodiversity and improving local air quality. Despite the several positive aspects, the CDM projects may also have broader social, ecological, economical, and environmental issues, which need to be tackled wisely.

Investing in sinks today according to the CDM by afforestation and reforestation may not help in cutting a lot of emissions tomorrow. Putting effort and emphasis on increasing sinks through afforestation and reforestation, instead of curbing GHG emissions, will add more carbon to the active carbon pool. In the race to increase sinks, the CDM may attract unfavourable technologies that will impose adverse impacts on the natural ecosystem and biodiversity. Looking into the commercial aspects of CDM for the carbon price, and being a funding agency, the developed country party may dictate to the developing country for genetically engineered, mono-culture, fast-growing species (Sawyer, Meinshausen, and Guilbeault 2003). This will not only negatively affect native biodiversity, but also put tremendous pressure on the indigenous agro-forestry system. Although plantation and afforestation help in reducing GHG emissions, no man-made plantation can create the natural interrelationships and habitats of biodiversity. Further, CDM projects, in the form of afforestation projects, may remove legal and customary land- and resource-use rights of local communities or indigenous peoples over forest land.

Past experience indicates that the advent of market forces in plant and animal species generally imposes pressure on existing biodiversity. Whenever traditional knowledge about the use of biodiversity is brought into the market, it generally makes negative impact on the population of such useful flora and fauna (Kala 2010b). Market forces have superseded traditional norms of conservation (Kala 2010c). The presumptive market in plantation projects, especially for CDM projects, is, other way round, an alarm call for the existence of native biodiversity if concerted efforts are not made for its conservation. Plantation not only alters soil microbes, but also changes the environment for native species. The monoculture plantation of eucalyptus tree in areas of Central India has influenced the groundwater table as well as the native biodiversity. Despite the many uncomfortable questions that it brings up, CDM creates numerous opportunities, including additional revenue streams, better practices, more energy-efficient technologies, and greater environmental awareness. A CDM project should be planned keeping in mind the sustainable development strategy and the conservation of native biodiversity.

To meet the goal of biodiversity conservation, all the scientific, commercial, and political forces need to work together, and conservation organizations should enhance their skills to integrate natural and social concerns into their approach. A multidisciplinary approach with a focused aim may be effective in achieving the desired targets, as evident from the glimpses of past research carried out in isolation. Although the challenges are many, if concerted efforts are made, then it is not too late to mitigate the climate change impacts on biodiversity by lowering the level of greenhouse gas emissions in the atmosphere.

REFERENCES

Balmford, A., L. Bennun, B. T. Brink, D. Cooper, I. M. Côté, P. Crane, A. Dobson, N. Dudley, I. Dutton, R. E. Green, R. D. Gregory, J. Harrison, E. T. Kennedy, C. Kremen, N. Leader-Williams, T. E. Lovejo, G. Mace, R. May, P. Mayaux, P. Morling, J. Phillips, K. Redford, T. H. Ricketts, J. P. Rodríguez, M. Sanjayan, P. J. Schei, A. S. van Jaarsveld, and B. A. Walther. 2005. The convention on biological diversity's 2010 target. *Science* 307:212–213

Bawa, K. S., R. Seidler, and P. H. Raven. 2004. Reconciling conservation paradigms. *Conservation Biology* 18(4):859–860

Defries, R., A. Hansen, A. C. Newton, and M. C. Hansen. 2005. Increasing isolation of protected areas in tropical forests over the past 20 years. *Ecological Applications* 15(1):19–26

Gairola, S., N. M. Shariff, A. Bhatt, and C. P. Kala. 2010. Influence of climate change on production of secondary chemicals in high altitude medicinal plants: issues need immediate attention. *Journal of Medicinal Plants Research* 4(18):1825–1829

Graves, J., and D. Reavey. 1996. *Global Environmental Change: Plants, Animals and Communities.* Essex: Longman Group

Hays, G. C., A. J. Richardson, and C. Robinson. 2005. Climate change and marine plankton. *Trends in Ecology and Evolution* 20(6):337–344

Kaiser, J. 2003. An intimate knowledge of trees. *Science* 300:566–567

Kala, C. P. 2005. Assessment of biodiversity loss. *Current Science* 89(7):1067

Kala, C. P. 2010a. Status of an indigenous agro-forestry system in changing climate: a case study of the middle Himalayan region of Tehri Garhwal, India. *Journal of Forest Science* 56(8):373–380

Kala, C. P. 2010b. *Medicinal Plants of Uttarakhand: Diversity, Livelihood and Conservation.* Delhi: Biotech Books

Kala, C. P. 2010c. Ethnobotanical and ecological approaches for conservation of medicinal and aromatic plants. *Acta Horticulturae* 860:19–26

Lomborg, B. 2001. *The Skeptical Environmentalist.* Cambridge: Cambridge University Press

Marquis, R. J. 2004. Herbivores rule. *Science* 305:619–621

Mawdsley, J. R., R. O'Malley, and D. S. Ojima. 2009. A review of climate-change adaptation strategies for wildlife management and biodiversity conservation. *Conservation Biology* 23(5):1080–1089

Moore, P. D. 2005. Parasite rattles diversity's cage. *Nature* 433:119

Naeem, S., and S. Li. 1997. Biodiversity enhances ecosystem reliability. *Nature* 390:507–509

Naeem, S., L. J. Thompson, S. P. Lawler, J. H. Lawton, and R. M. Woodfin. 1994. Declining biodiversity can alter the performance of ecosystems. *Nature* 368:734–737

Nathan, R., N. Horvitz, Y. He, A. Kuparinen, F. M. Schurr, and G. G. Katul. 2011. Spread of North American wind-dispersed trees in future environments. *Ecology Letters* 14(3):211–219

Nemani, R. R., C. D. Keeling, H. Hashimoto, W. M. Jolly, S. C. Piper, C. J. Tucker, R. B. Myneni, and S. W. Running. 2003. Earth grows greener. *Science* 300:1560–1562

Parmesan, C. 2006. Ecological and evolutionary responses to recent climate change. *Annual Review of Ecology, Evolution and Systematics* 37:637–669

Parmesan, C., and G. Yohe. 2003. A globally coherent fingerprint of climate change impacts across natural systems. *Nature* 421:37–42

Redford, K. H., P. Coppolillo, E. W. Sanderson, G. A. B. Da Fonseca, E. Dinerstein, C. Groves, G. Mace, S. Maginnis, R. A. Mittermeier, R. Noss, D. Olson, J. G. Robinson, A. Vedder, and M. Wright. 2003. Mapping the conservation landscape. *Conservation Biology* 17(1):116–131

Rezende, E. L., J. Balanyà, F. Rodríguez-Trelles, C. Rego, I. Fragata, M. Matos, L. Serra, and M. Santos. 2010. Climate change and chromosomal inversions in *Drosophila subobscura. Climate Research* 43:103–114

Rodrigues, A. S. L., S. J. Andelman, M. I. Bakarr, L. Boitani, T. M. Brooks, R. M. Cowling, L. D. C. Fishpool, G. A. B. Fonseca, K. J. Gaston, M. Hoffmann, J. S. Long, P. A. Marquet, J. D. Pilgrim, R. L. Pressey, J. Schipper, W. Sechrest, S. N. Stuart, L. G. Underhill, R. W. Waller, M. E. J. Watts, and X. Yan. 2004. Effectiveness of the global protected area network in representing species diversity. *Nature* 428:640–643

Sala, O. E., F. S. Chapin, J. J. Armesto, E. Berlow, J. Bloomfield, R. Dirzo, E. Huber-Sanwald, L. F. Huenneke, R. B. Jackson, A. Kinzig, R. Leemans, D. M. Lodge, H. A. Mooney, M. Oesterheld, N. L. Poff, M. T. Sykes, B. H. Walker, M. Walker and D. H. Wall. 2000. Global biodiversity scenarios for the year 2100. *Science* 287:1770–1774

Sawyer, S., M. Meinshausen, and S. Guilbeault. 2003. Sinks in the CDM: After climate, will biodiversity go down the drain? Greenpeace. Details available at <www.greenpeace.org/international/Global/international/planet-2/report/2006/3/sinks-in-the-cdm-after-the-cl.pdf>

Storch, D., P. A. Marguet, and K. J. Gaston. 2005. Untangling an entangled bank. *Science* 307:684–686

Walter, G. R., E. Post, P. Convey, A. Menzel, C. Parmesan, T. J. C. Beebee, J. M. Fromentin, O. Hoegh-Guldberg, and F. Bairlein. 2002. Ecological responses to recent climate change. *Nature* 416:389–395

CHAPTER 17

Role of Forests and Associated Traditional Knowledge of Local Communities in Climate Change Adaptation

Chandra Shekhar Silori

INTRODUCTION

The debate on the role of forests in addressing climate change is most prominent in the development sector. The debate is particularly significant for the livelihood security of forest-dependent communities and their vulnerability to climate change. The number of such communities is estimated as 450 million people in Asia Pacific alone. These people rely on forests to different degrees and in different ways, based on various factors. Generally, in developing countries farmers rely on forests to collect wood and small timber for the market, while others earn additional income through the sale of non-timber forest products (NTFPs). Many products taken from the forest never make it to the market but are consumed by the local communities for their subsistence needs. For such households and communities, forests provide an important safety net during hard times, which include poor harvests, economic crises, and natural disasters.

Climate change is expected to have a huge impact on forests and the livelihoods of forest-dependent people. The *Fourth Assessment*

Report (*AR4*) (2007) of the Intergovernmental Panel on Climate Change (IPCC) projects with high confidence that change in the frequency and severity of extreme climate events will have significant impact on forests and agricultural production and, therefore, on food security and poverty. It also projects that water resources will become increasingly scarce and the level of environmentally induced migration will increase. The *Human Development Report* of the United Nations Development Programme (UNDP) shows that climate change is not just a future scenario; increased exposure to droughts, floods, and storms is already destroying opportunity and reinforcing inequality (UNDP 2007). Thus, the phenomenon of climate change has the potential to dramatically affect the lives of present as well as future generations.

POTENTIAL IMPACTS OF CLIMATE CHANGE ON FORESTS AND FOREST-DEPENDENT COMMUNITIES

The impact of climate change will vary from one region to another, and may not unfold exactly as predicted. However, despite our limited knowledge on the magnitude and direction of change, it is nearly certain that changes will occur, and that some changes have already started showing their impact on the human society. The IPCC's *AR4* highlighted the potential impact of climate change on forest ecosystems. Increased surface temperature of more than 2°C above pre-industrial levels are likely to result in substantial changes in the structure and function of forest ecosystems (Midgley, Price, Fischlin, *et al.* 2007). New findings indicate that the negative impacts of climate change on forests may be more severe than previously believed and that the potential positive impacts have been overestimated. The negative impacts will include biome redistribution, increased forest fire occurrence, and accelerated infestation of weeds and other pests.

The poor and marginal communities across the globe are predicted to be more vulnerable to climate change in general, and particularly to the changes in the forest ecosystems. Therefore, the adverse impact is more severely felt by such communities who live in marginal lands and whose livelihoods are largely dependent on natural resources.

Should these changes take place, the result will be an unprecedented level of conflict over the management of already scarce and often degraded forest resources. This stands to be particularly true where user rights are poorly defined and where governance systems are weak. At a regional scale, climate change is projected to impinge the most on livelihoods of human society of developing countries and regions that are exposed to high climatic variability (UNFCCC 2007). Countries located in the Arctic, small islands, arid and semi-arid regions, high altitudes, and in the developing world are highly vulnerable to climate change.

The existing vulnerability of many local communities will confer on them an additional vulnerability: without supportive infrastructure, effective political processes, or other sources of sustenance, such communities will be less able to respond to any changes that do unfold (Seppala, Alexander, and Katila 2009). The IPCC makes it clear that although climate change is global, 'local efforts can help maintain and enhance resilience and limit some of the long-term damages from climate change' (2007). These local efforts will need to be multi-faceted, including elements of resistance and resilience. Given the uncertainty inherent in climate change modelling, any adaptation or local effort must necessarily include a general response plan, one that looks at what can be done immediately to encourage and facilitate adaptation to change.

SUSTAINABLE FOREST MANAGEMENT AND CLIMATE CHANGE ADAPTATION

In light of the predicted changes in climate, the need is more than ever for planned and anticipatory adaptation (Buck, Katila, and Sepala 2009; IPCC 2007; ITTO 2009). Recent studies strongly point to the potential role of forests in climate change adaptation and mitigation. Indeed, Fischlin, Midgley, Price, *et al.* (2007) argue that forest management may be one of the only tools that can effectively influence how climate change affects forests. Sustainable forest management supports several environmental services: it protects watersheds and provides a buffer against landslides, tsunamis, and natural disasters. In rural areas, forests are also essential sources of livelihood for local communities and indigenous people.

Community forestry management (CFM) is a tool that is flexible and well-tested to build adaptation capacity and resilience in forest-dependent communities to climate change. Good forest management by local communities and indigenous people is exactly the type of broad response that will encourage and facilitate other forms of adaptation. As such, it is pivotal to any adaptation plan in forested landscapes. Further, there is a growing consensus that community forestry and similar forms of devolved forest management will be the key to climate change adaptation. The ITTO (2009) has made a strong case for community forestry, arguing that 'countries will require flexible and equitable decision-making processes at local and national levels that allow rapid and adaptable management responses'. Support for community forestry derives from the fact that it is tested and proven: it is an effective means of addressing pressing social, economic, and environmental issues such as rural transformation and urbanization, market changes, and water scarcity. Support also derives from the growing body of research showing that communities with community-based resource management are likely to be more resilient to climate change (Tompkins and Adger 2004; Silori, Roongwong, and Soontornwong 2009).

Community forestry management has evolved from isolated pilot initiatives in a few countries like India, Nepal, and the Philippines in the early 1970s to a well-established, institutionalized process in more than 100 countries (Malla 2000; Gilmour, Malla, and Nurse 2004). Many countries mainstreaming CFM in national forest programmes and establishing links between 800 million (67%) of the world's poorest people and forests are indications of its growing importance and acceptance to address widespread forest loss, the consequent environmental degradation, and the negative impact of climate change. According to a recent report by the Rights and Resources Initiative, there are around 17.8 million hectares (Mha) designated for use by local communities and indigenous people in Asia, besides, nearly 146.6 Mha owned by local communities and indigenous people, and the forest areas under community management in CFM pioneering countries such as Nepal (1.25 Mha) and the Philippines

(5.8 Mha). Areas managed under formal arrangements continue to grow. The growing number of local forest managers requires support so that the benefits derived can be secured and increased. This is of particular importance, as the livelihoods of local communities and indigenous people will continue to be affected by climate change.

INDIGENOUS KNOWLEDGE ON COPING STRATEGIES

Another important dimension of CFM is forest-related traditional and indigenous knowledge, which provides ample opportunities to build the resilience of local communities to climate change. Unfortunately, traditional knowledge is only rarely considered in academic, policy-related, and public discourse on climate change, even though it will be greatly affected by impending changes (Macchi 2008). In developing countries, indigenous people are not only keen observers of climate change but also try actively to adapt to the changing conditions. Therefore, they are vital and active parts of many ecosystems and may help to enhance their resilience. In addition, they interpret and react to climate change impact in creative ways, drawing on traditional knowledge as well as new technologies to find solutions, which may help society at large to cope with impending changes (Byg and Salick 2007; Silori 2008).

Most published reports on indigenous observation of climate change have come from Arctic regions, perhaps because of the strong co-operation between scientists and the indigenous people (Berkes and Reidlinger 2000; Berkes and Jolly 2001). However, there are innumerable examples of indigenous coping strategies to the uncertainties and adaptation to climate change from other parts of the world as well. In Asia and the Pacific, most communities being dependent on natural and cultivated resources, the traditional knowledge resources base on coping strategies is enriched with a number of indigenously developed methods/technologies to counter the uncertainties of climate change. Forest lands conserved and protected by Van Panchayats in Uttarakhand, India (Box 1) and community forest management by many indigenous communities in many parts of the developing world are a few such examples.

Box 1 Van Panchayats of Uttarakhand, India

Van Panchayats (VP) are unique institutions, characteristic of Uttarakhand state in north India, formed for organized utilization and protection of forests and other related natural resources by local communities dependent on them. VP were born out of conflicts and compromises that followed settlements and reservations in the hills at the turn of the last century. They are locally elected bodies or voluntary groups of local people that govern the nearby forests to fulfil the needs of the local people for forest produce in a sustainable and equitable manner. Thus, VP are a kind of local government. They have been in existence for over 75 years. However, the association of the people of Uttarakhand with forests and their history of protecting them is longer and eventful. There are 6777 VP in the state, covering an area of 5241.08 km^2.

The existence of sacred groves in almost all the countries of South Asia is an example of self-regulation by local communities against cutting/damaging trees, grazing animals, and removal of biomass in any form (Malhotra, Gokhale, Chatterjee, *et al.* 2007). These regulations are part of the efforts to maintain the health of the ecosystem and cope with unpredictable climates. Traditionally, cultivation of medicinal and wild edible herbs in homestead gardens in many hilly regions of India, Nepal, and Bhutan is another important practice that not only meets the health requirements of local people but also helps them earn from the sale of such plants/products and reduce their vulnerability by having additional resources to survive during hard times.

Traditional agro-forestry models are other examples that have proved useful as support to ensure food security, income, and meet biomass needs during adverse situations. In Sri Lanka, coconut is the most widely planted agro-based industrial tree, covering 25% (419 200 ha) of the total cultivated area in the country. On most coconut estates, a variety of different crops are cultivated, including food crops (such as tubers, cereals, legumes, and fruits), spices and condiments (such as areca nut, betel leaves, chillies, ginger, and turmeric), and cash crops (black pepper, cacao, cinnamon, cloves, coffee, and nutmeg). On some plantations, pasture grass is grown in between coconut trees to allow cattle to graze.

Similar practices are reported from eastern Himalayan countries (Box 2). Communities in the Indian Himalayas face erratic rainfall during spring and summer. This indicates shortening of seasons (2–5 months) and overexploited, sparse natural vegetation, which are unable to meet the energy and fodder needs of the local communities. Farmers in such areas have developed traditional agro-forestry practices to ensure food security and additional income generation. For example, they cultivate bamboo groves in all agricultural holdings along streams and irrigation/drainage channels in high rainfall areas of Sikkim, India. On maturity, such bamboo groves provide multiple benefits to local communities. Bamboo is extensively used in building small farmhouses, goat sheds, pig enclosures, baskets, and string making. Large bamboo pipes are used as water conveyers in the farm irrigation/drainage system. Bamboo leaves are excellent

Box 2 Large cardamom-based agro-forestry in Eastern Himalayas

Large cardamom-based agro-forestry has been a boon to the people of Sikkim for a very long time. It is widely cultivated under the nitrogen-fixing Himalayan alder *(Alnus nepalensis)*, a practice modified by people to maintain soil fertility and increase productivity. The fruit produced are used widely as spices/condiments and contain about 3% essential oils rich in cineole. For the past three decades, this traditional agro-forestry system has become so popular that it was scaled up through community exchange in the neighbouring countries of Nepal and Bhutan. It meets the adaptive requirements of mountain specificities such as the following.

- Inaccessibility—by producing marketable but non-perishable products;
- Marginality—by growing it on marginal lands belonging to the poor and indigenous peoples;
- Fragility—by providing forest cover and a perennial agricultural cash crop; and
- Opportunities for increased income—because it is a niche cash crop.

The economic benefits derived from large cardamom-based agro-forestry in Sikkim reported an increased gross income, from $1.9 million in 1975/76 to $6.4 million in 1995/96. This large cardamom-based agro-forestry is a perfect example of mountain landscape management in which both the ecological and the economic aspirations of local, regional, and international communities are achieved (Sharma 2006).

winter fodder for goats, and the stumps protect water channels from erosion.

DISCUSSION AND CONCLUSION

While it is satisfying to see the growing recognition of the forestry sector in addressing climate change, there are a number of issues that need further clarity to strengthen the role of forestry in countering the impact of climate change. Despite growing acreage and the number of participating communities and countries in devolved forest management all over the globe, there are several issues that continue to be challenging. The governance, rights, and tenure security of local communities and indigenous people over the forest and other common property land resources and benefit-sharing continue to challenge the execution abilities of practitioners and policymakers to build the adaptation abilities of the forest-dependent communities to climate change. On the other hand, standard operating procedures in forest bureaucracies need to change for adaptation approaches to be useful and successful.

Brockhaus, Colfer, Kanninen, *et al.* (2008) stress two key changes that are required. Firstly, forest-dependent people need to be recognized, drawn into the discourse, and should contribute to the official decision making. This requires a dramatic change in the attitude of forestry personnel and strengthening feedback mechanism within forestry related bureaucracies. To enhance environmental services and, thus, the adaptation strategies of forest-dependent communities, biodiversity conservation needs to be given adequate attention in community forest areas. Through such measures, the livelihood products provided by forests will be enhanced and the safety net for marginal and landless communities will be broadened (Noss 2001; Bergeron, Drever, Flannigan, *et al.* 2006).

There are various ways through which biodiversity conservation can be enhanced and promoted for the betterment of local communities (Boven and Morohashi 2002; Byg and Salick 2007; Roberts 2009; Guariguata, Garcia Fernandez, Sheil, *et al.* 2010). The role of local communities and indigenous people in all these measures is going to be crucial. The local communities have developed traditional knowledge practices and institutions over generations, and have continuously adapted to the changing environmental conditions within the forest ecosystems. Therefore, such knowledge needs to be mainstreamed in developing policies and relevant tools to adapt to climate change. Since such knowledge cuts across various other natural resources—including water resource management, soil conservation practices, and flood management—robust management strategies and intersectional coordination need to be created and enhanced at national and local levels while developing adaptation strategies. All these initiatives need to be supported by a well-developed and flexible monitoring and evaluation system to constantly learn and improve adaptation strategies to counter the impact of climate change.

REFERENCES

Bergeron, Y., C. R. Drever, M. D. Flannigan, C. Messier, and G. Peterson. 2006. Can forest management based on natural disturbances maintain ecological resilience? *Canadian Journal of Forest Research* 36:2285–2299

Berkes, F., and D. Jolly. 2001. Adapting to climate change: social-ecological resilience in a Canadian western Arctic community. *Conservation Ecology* 5(2):18. Details available at <www.consecol.org/vol5/iss2/art18>

Berkes, F., and D. Riedlinger. 2000. Contributions of traditional knowledge to understanding climate change in the Canadian Arctic. Details available at <www.arcus.org/award/PDF/ 5th_PDF/Riedlinger_Abstract.pdf>

Boven, K., and J. Morohashi (ed.). 2002. *Best Practice using Indigenous Knowledge*. The Hague, Nuffic and Paris: UNESCO/MOST. Details available at <www.unesco.org/most/Bpikpub2.pdf>

Brockhaus, M., C. J. P. Colfer, M. Kanninen, B. Locatelli, D. Murdiyarso, and H. Santoso. 2008. Facing an uncertain future: how forests and people can adapt to climate change. [Forest Perspectives No. 5, Bogor, Indonesia, CIFOR]

Buck, A., P. Katila, and R. Sepala (ed.). 2009. *Adaptation of Forests and People to Climate Change: A Global Assessment Report*. World Series, vol. 22. Helsinki: International Union of Forest Research Organization

Byg, A., and J. Salick (ed.). 2007. *Indigenous People and Climate Change*. Oxford, UK: A Tyndall Centre Publication

Cronkleton, P., C. Garcia-Fernandez, M. Guriaguata, C. Herrero-Jauregui, V. Ingram, R. Nasi, and D. Sheil. 2010. Compatibility of timber and non-timber forest product management in natural tropical forests: perspective, challenges and opportunities. *Forest Ecology and Management* 259(3): 237–245

Fischlin, A., G. F. Midgley, J. T. Price, R. Leemans, B. Gopal, C. Turley, M. D. A. Rounsevell, O. P. Dube, J. Tarazona, and A. A. Velichko. 2007. Ecosystems, their properties, goods and services. In *Climate Change 2007: Impacts, Adaptation and Vulnerability, Contribution of Working Group II to the Fourth Assessment Report of the Intergovernmental Panel on Climate Change*, edited by M. L. Parry, O. F. Canziani, J. P. Palutikof, P. J. van der Linden, and C. E. Hanson, pp. 211–272. Cambridge, UK: Cambridge University Press

Gilmour, D., Y. Malla, and M. Nurse. 2004. *Linkage Between Community Forestry and Poverty*. Bangkok: Regional Community Forestry Training Center for Asia and the Pacific (RECOFTC)

Guriaguata, M., C. Garcia-Fernandez, D. Sheil, R. Nasi, C. Herrero-Jauregui, P. Cronkleton, and V. Ingram. 2010. Compatibility of timber and non-timber forest product management in natural tropical forests: perspective, challenges and opportunities. *Forest Ecology and Management* 259(3): 237–245

ITTO (International Tropical Timber Organization). 2009. Partnership to conserve tropical forests launched. Details available at <www.itto.int/direct/topics/topics_pdf_download/topics_id=2130&no=0>

IPCC (International Panel on Climate Change). 2007. *Summary for Policymakers. Climate Change 2007: the Physical Science Basis. Contribution of Working Group I to the Fourth Assessment Report of the Intergovernmental Panel on Climate Change*, edited by S. Solomon, D. Qin, M. Manning, Z. Chen, M. Marquis, K. B. Averyt, M. Tignor, and H. L. Miller. Cambridge, UK, and New York, USA: Cambridge University Press

Macchi, M. 2008. Indigenous and traditional peoples and climate change. Gland: International Union for Conservation of Nature [Issues paper]

Malla, Y. B. 2000. Impact of community forestry policy on rural livelihoods and food security in Nepal. *Unasylva* 202:37–45

Malhotra, K. C., Y. Gokhale, S. Chatterjee, and S. Srivastava. 2007. *Sacred Groves in India*. New Delhi: Aryan Books International

Noss, R. F. 2001. Beyond Kyoto: forest management in a time of rapid climate change. *Conservation Biology* 15:578–590

Roberts, G. 2009. Current adaptation measures and policies. In *Adaptation of Forest and People to Climate Change: A Global Assessment Report*, edited by R. Sepalla, A. Buck, and P. Katila, World Series, vol. 22, pp. 123–133. Vienna: International Union of Forest Research Organization

Seppala, R., B. Alexander, and P. Katila. 2009. *Adaptation of Forests and People to Climate Change: A Global Assessment Report*, World Series, vol. 22. Vienna: International Union of Forest Research Organization

Sharma, R. 2006. Traditional agroforestry and a safer mountain habitat. Details available at <www.icimod.org/-home/uploads/newsletter/nl50/traditional.pdf>

Silori, C. S. 2008. Climate change: indigenous coping strategies in South Asia. *South Asian Journal* 21:8–23

Silori, C. S., A. Roongwong, and S. Soontornwong. 2009. Links between conservation and livelihood security in practice: a case study of community conservation of mangrove forests in Pred Nai, Thailand. Paper presented in the *International Workshop on Community Forestry* held in Pokhara, Nepal, 15–18 September 2009

Tompkins, E. L., and W. N. Adger. 2004. Does adaptive management of natural resources enhance resilience to climate change? *Ecology and Society* 9(2):10. Details available at <www.ecologyandsociety.org/vol9/iss2/art10>

UNDP (United National Development Programme). 2007. Fighting climate change: human solidarity in a divided world. In *Human Development Report 2007/2008*. Details available at <www.hdr.undp.org/en/media/hdr_20072008_en_complete.pdf>

UNFCCC (United Nations Framework Convention on Climate Change). 2007. *Climate Change: Impacts, vulnerabilities and adaptation in developing countries*. Bonn: UNFCCC Secretariat

CHAPTER 18

Adaptation and Mitigation Strategies of Climate Change for Sustainable Agriculture

Arvind Bijalwan

INTRODUCTION

The agriculture sector—the lifeline of India—is concerned not only with the growing of grains, cereals, pulses, oil seeds, fruits, vegetables, spices, and condiments, but also with associated livelihood activities such as animal husbandry, horticulture, fisheries, and aquaculture. In India, the total cultivated area is approximately 143 million hectares (Mha), out of which about 50% is rainfed agriculture, which is highly sensitive to climate. The agricultural sector contributes to approximately 15.7% of India's GDP (2008–09) and 10.23% (provisional) of total exports, besides providing employment to around 58.2% of the workforce (DAC 2010) of the country. This indicates the role of the agriculture sector in safeguarding national food security, a crucial issue in the era of global climate change. The agriculture sector is vulnerable to changing climatic conditions, and is likely to subsequently affect food and water security (Swaminathan 2010). This is further complicated by the fact that nearly 120.72 Mha are affected by various kinds of land degradation such as water and wind erosion, water logging, alkalinity/salinity, soil acidity, as reported by the National Rainfed Area Authority (NRAA). Frequent droughts, floods, and climate change have negatively affected soil fertility and land degradation, thereby affecting food grain production in the country (DAC 2010).

The diversity of geographical and climatic conditions makes India suitable for the cultivation of a variety of agricultural crops, including wheat, paddy, maize, and jowar among cereals; gram, tur, urad, and moong among pulses, and soya bean, groundnut, and mustard among oilseeds. Most coarse grains (millets) are resistant to adverse climatic conditions and sustain the lives of rural people in extreme climatic conditions. However, the area under coarse grain cultivation is shrinking day by day, and is being replaced by wheat and other commercial crops. Coarse grains are a boon for dry-land agriculture.

BRIEF OVERVIEW OF CLIMATE CHANGE IN INDIA

During the past century, the average global air temperature near the Earth's surface has been estimated to increase at the rate of 0.74 ± 0.18ºC (1.33 ± 0.32°F) (IPCC 2007a). The Indian climate change projections for 2030 derived by the Indian Network for Climate Change Assessment (INCCA) in its second report (2010b) elaborated that an all-round warming over the Indian subcontinent is associated with increasing green house gases (GHG) concentrations. The annual mean surface air temperature is projected to rise by 1.7ºC to 2.0ºC in the 2030s, with 1970 as the base reference year. Seasons may be warmer by around 2.0ºC towards the 2030s, whereas the variability of seasonal mean temperature may be more in winter months (INCCA 2010a). Similarly, a small increase in annual precipitation is predicted by the 2030s, with respect to the baseline of the 1970s; further, the projection of frequency of rainy days is likely to decrease in most parts of the country (INCCA 2010a). Although there are many uncertainties in the projections of impact and vulnerability of climate change due to several complexities, the potential causes and their mitigation strategies need to be identified and addressed. In the agricultural sector, the basic fact of reduction in productivity, shift in sowing and harvesting time of crops, heat stress to livestock, and migration of fish to low-temperature zones are some major concerns that need proper investigation and understanding. This chapter reviews various adaptation and mitigation strategies for sustainable and resilient agriculture with respect to climate change.

LINKAGES BETWEEN AGRICULTURE AND CLIMATE CHANGE

The amount of GHG keeps increasing in the atmosphere. According to INCCA's assessment for India in May 2010, the GHG emissions are mainly from the energy, industry, agriculture, and waste sectors. They accounted for 58%, 22%, 17%, and 3% respectively, of the net carbon dioxide equivalent (CO_2e) emission in 2007 (INCCA 2010b). The agriculture sector alone emitted 334.41 MT CO_2 in 2007 from fermentation in livestock manure, paddy cultivation, agricultural soils, and burning of crop residues. The INCCA 2010 report further clarified that, due to enteric fermentation, the Indian livestock sector released 212.1 MT CO_2e (10.1 MT CH_4), which constituted 63.4% of the total GHG emissions (CO_2e) from the agriculture sector in India. The estimates cover all livestock (cattle, buffalo, sheep, goats, poultry, donkeys, camels, horses). Manure management is estimated to emit 2.44 MT CO_2e. Similarly, paddy fields are also considered to be a major source of CH_4 emission, which is 21 times more harmful than CO_2 (INCCA 2010b). Rice cultivation emitted 69.87 MT CO_2e or 3.27 MT CH_4. The emission from the rice field mainly depends on water management practices. Rice fields irrigated through flooding and deep water emit maximum CH_4, while upland rice is generally in the category of zero emitters. There is a traditional practice of agricultural soils and fields burning crop residue in India—this also augments the amount of GHGs in the atmosphere. Agricultural soils are a source of nitrous oxide (N_2O), mainly due to the application of nitrogenous fertilizers in the field, whereas burning of crop residues leads to the emission of a number of gases and pollutants. Hence, both agricultural soils and crop residue burning account for the emission of 50.00 MT CO_2e (INCCA 2010b).

IMPACTS OF CLIMATE CHANGE ON AGRICULTURE

Any change in climate may disturb and deteriorate the agriculture and livestock sectors. The impact of climate change on Indian agriculture could be devastating, especially for vulnerable crops, which may subsequently influence the future food security of the country. Climate change may lead to a decline in crop yield, soil health, crops sequences, and practices followed by farmers in different agro-climatic regions. Various regions of India have been experiencing such effects in the aberration of temperature—from frost injuries to heat stress and changes in rainfall pattern—resulting in an imbalance in the physiological and biochemical activities of crops, which leads to reduction in crop yield. In the middle Himalayan region, there has been a noticeable impact of climate change on the traditional agricultural pattern, particularly due to erratic rainfall and rise in temperature. It was reported that a prolonged dry spell in winter has delayed the time for sowing wheat; reduction of yield was also noticed in traditional vegetable crops such as peas, potato, ginger, and beans. At the same time, the crop maturation period has advanced by 1–2 months (Joshi and Joshi 2011). The size of Gaderi or Arabi (*Colocasia esculenta*), a type of yam usually grown in the hills, has reduced from 9–10 inches to 4–6 inches and its weight has reduced from 2–3 kg to 0.25–0.5 kg in recent years. The local people attributed this mainly to the changing weather patterns (Joshi and Joshi 2011). The unpredictable changes in the onset of the monsoon are likely to affect crop sequence and sowing time, which ultimately lowers crop yield. In general, the major meteorological parameters—such as change in temperature, alteration in rainfall pattern, and their impact on soil moisture level—may place conditions under high stress and affect rural livelihood. Although the scientific confirmation in relation to climate change warning to the agricultural segment is explicit, the complex interactions and opinions are uncertain.

CLIMATE CHANGE CONCERN IN HORTICULTURE

Horticultural crops, an allied sector of agriculture, include vegetables and perennial fruit crops. They have immense scope for the development of small and marginal cultivators. This sector yields regular income, due to high productivity and valuable produce, compared to food grains. Water is a major constraint for horticultural crops across the country; excess deficiency of water may create a drought situation. As fruit trees are perennial in nature, the impact of drought in them is recognized only in the late stage, making it difficult for farmers to adopt any precautionary measures in time. The rise in temperature, particularly in fruit crops, can lead to early ripening of fruits and thereby increase the possibility of fruit fall before time. Rise in temperature can also lead to earlier pest emergence, which increases the vulnerability of plants to aphid attacks. A series of diseases has been observed in fruit plants due to change in climatic parameters, particularly wilting and top dying. Abnormal storms and hailstorms damage horticultural crops during flowering time to a large extent. Heavy rains also cause physical damage to plants and fruits when they occur during the fruit-bearing stage.

The increased heat stress may change the fruit size and pollination pattern in some fruit crops. High temperature can also reduce tuber initiation in potatoes. In fruits like apples, the anthocyanin is responsible for colour development which is also reduced at higher temperature. Capsicum is likely to face similar problems. Most fruit crops require cross pollination, and increase in temperature can adversely affect the synchronization of flowering, resulting in a reduction in yield. The rise in temperature may increase disease and pest infestation, which directly impacts the growth rate of crops. There may be chances of a hike in charges of cold storage (post-harvesting) due to the rise in temperature in most of the vegetables and perishable crops. In addition, the large floriculture sector of India may also be influenced by the odd performances in climatic parameters. It has been observed that the gradual melting of ice caps in the Himalayan region is likely to reduce the chilling requirement for flowers like tulips, orchids, rhododendrons, and alstromerea, and some of them are likely to fail to bloom (Sharma and Roy 2010). Commercial production of flowers, particularly grown under open field conditions, may be severely affected, leading to poor flowering, improper floral development, and colour, besides reduction in flower size and a shorter blooming period (Sharma and Roy 2010).

ANIMAL HUSBANDRY AND CLIMATE CHANGE

As far as animal husbandry and climatic change in India is concerned, the heat tolerance limit and reproductive behaviour of animals, particularly buffaloes, are matters of concern. The rise in temperature may lower the conception rate and fertility in animals, while creating favourable conditions for pathogens to reproduce, leading to different diseases in livestock. Crossbred cattle are more productive but the serious concern is about the acclimatization of cross bred in different local climatic situations. The increase in livestock population is also the source of more CH_4 emission. Another serious concern is that abrupt climatic conditions can pose a threat to the availability of feed, fodder, and a proper, balanced diet; this ultimately imposes severe threats to the survival of animals. Abnormal climatic conditions create malformed livestock and make them more susceptible to diseases and such unhealthy animals are a major source of CH_4 liberation.

CLIMATE CHANGE CONCERN IN FISHERIES

Inland fisheries of the Indian subcontinent have been affected due to the rise in temperature over the last few decades. It has forced fish communities to move upward (to colder regions). This has been reported in River Ganges, where a few fish species that were commonly found in the foothills are now reported from middle and upper stretches of the river (Sinha, De, and Jha 1998). A change in atmospheric temperature has been observed. For example, if there is a delay in monsoons, it leads to frequent droughts and floods. Furthermore, the unsympathetic storms in the coastal regions and the recession in Himalayan glaciers are witnesses to the changes in bio-physical parameters. The average rainfall has decreased; consequently, the climate change impact is being felt on the temperature of the inland water bodies and on the breeding behaviour of fish (Mohanty, Mohanty, Sahoo, *et al.* 2010). Inland water bodies including the major river system like the River Ganges provide a life support system to all human beings, especially the fishing community, which depends on these water bodies for its livelihood. There are reports that annual mean minimum water temperature in the upper cold-water stretch of the River Ganges in Haridwar has increased by 1.5°C (from 13°C during 1970–86 to 14.5°C during 1987–2003). This change in temperature has changed the biogeographical distribution of the Gangetic fish fauna (Vass, Das, Srivastava, *et al.* 2009).

ADAPTATION TO AND MITIGATION OF CLIMATE CHANGE IN THE AGRICULTURE SECTOR

Adaptation to climate change is defined as an adjustment in ecological, social, or economic systems, in response to observed or expected changes in climatic stimuli and their effects and impact to alleviate adverse effects of change, or take advantage of new opportunities (IPCC 2007b). Adaptation is an active set of strategies and actions taken by people in response to, or in anticipation of, the change to enhance or maintain their well being. Hence, adaptation is a continuous stream of activities, actions, decisions, and attitudes that informs decisions about all aspects of life and that reflects existing social norms and processes (Daw, Adger, Brown, *et al.* 2009). Although mitigation is equally important, it is a long-term strategy and the real beneficiaries of mitigation will be the future generations. The choice between mitigation and adaptation strategies has spatial (geographic) and temporal (different generation) dimensions.

There are prime concerns of adaptation to and mitigation of climate change in the agricultural sector (crop husbandry, horticulture, animal husbandry, fisheries). Adaptation is much more important and the impact and benefit of adaptation measures are immediate. In India, the adaptive capacity of dry-land farmers, forest dwellers, fisherfolk, and nomadic shepherds is very low (Ravindranath and Sathye 2002). There are two main processes of adaptation: autonomous and planned. Adaptation that does not constitute a conscious response to climatic stimuli but, rather, is triggered by ecological changes in natural systems and by market or welfare changes in human system is referred to as autonomous or spontaneous adaptation (IPCC 2007b). Planned adaptation is the result of a deliberate policy decision, based on the awareness that conditions have changed or are about to change and that action is required to return to, maintain, or achieve a desired state (IPCC 2007b). There is an important approach to compare the cost and benefits of adaptation and mitigation strategies for creating equilibrium in society. Further, if adaptation to climate change could be carried out at negligible cost then it may be less expensive, at least in the short-term, than any alternate strategy (Sathaye, Shukla, and Ravindranath 2006). In fact, there are difficulties establishing the benefits of adaptation policies and consequently avoiding damages (Jempa and Munasinghe 1998).

SUGGESTIVE ADAPTATION AND MITIGATION STRATEGIES

Agriculture

It has been perceived in many parts of the country that there is an average shift in the onset of monsoon that completely upsets the cropping pattern and cycle, resulting in lowered agricultural productivity. The indicative adaptive measures such as late sowing crop varieties with short rotation and proper maturity time are important considerations, besides supporting dry farming as an optional and feasible adaptive measure in the Indian scenario. Erratic rainfall is another major concern. Despite the fact that there is ample rainfall, its irregular distribution severely affects the growth and development of agricultural crops and soil moisture. Mulching techniques (plastic mulching/bio- or organic mulching) are proposed as potential adaptation measures for maintaining soil moisture in agricultural fields. The application of surplus nitrogenous fertilizer is the prime source of N_2O, an important GHG from agricultural soil; therefore, promoting organic farming as a better option to reduce the emission of N_2O. Excess use of nitrogenous fertilizers also leads to soil hazards that disturb soil micro-fauna and, ultimately, the interrelationship of microbes-plant-climatic situation. As the availability of organic matter in farmers' fields makes the soil live and safeguards the moisture, appropriate application of organic farming (vermicomposting/farm yard manure [FYM]) in association with bio-fertilizers is also a viable option, especially for drought-prone areas. If there is a need to use inorganic fertilizers, it can be supplemented with organic farming for better and stable agriculture. The use of locally available biodegradable waste, bio-fertilizers, and bio-pesticide needs to be promoted for eco-friendly agriculture.

Hybrid and improved seed varieties may be promoted for cultivation but they need to be productive apart from being heat and drought tolerant. Cultivation of paddy (rice) is judged to be the major source of CH_4 emission; therefore, to minimize CH_4 emission, the System of Rice Intensification (SRI) in irrigated or low-land areas is an optional measure. The SRI technique of rice cultivation liberates low amounts of CH_4 and does not hamper the yield of paddy. As far as adaptation of agricultural crops in dry-land agriculture is concerned, coarse grain is the most suitable practice. As jowar, bajra, kodo, ragi, and other millets are resistant to adverse climatic situations, the area under coarse grain cultivation should be increased. Monoculture is generally not successful in changing climatic situations. Therefore, agriculture needs to be cultured with fruit trees and vegetables so that a mixed cropping practice may develop and it will lower the chances of completely diminishing the agricultural crops. As agricultural practices are the source and sink of carbon for short periods, trees may be planted with agriculture crops in agroforestry fashion (for example, agri-silviculture, agri-horticulture, agri-horti-silviculture, and silvipasture systems) for the long-term blockage of carbon in the trees. Besides, trees also help to conserve soil moisture and regulate the microclimatic situation.

Horticulture

Fruit trees, being perennial in nature, serve to sequester carbon for a long time. Horticultural production is a valuable sector for the Indian economy and water is an essential component for all horticultural components (vegetables and fruits). Increasing droughts and risks of climate change are some of the major threats to the sustainability of this sector. For better and efficient utilization of water, a large horticultural area should be put under micro-irrigation; this will protect the crop as well as enhance the productivity of the sector. Drip irrigation in horticulture has proved to be more efficient than the sprinkler for efficient utilization of water and the growth of fruit plants. Different mulching is also required for vegetable crops during the nursery stage to protect them from frost injuries, direct sunlight, and other adverse climatic abnormalities. Mulching also conserves the in-situ moisture content in the field. As, in general, indigenous vegetables are found to be more adaptive to the local climate and thereby perform better in quality and quantity, therefore their cultivation needs to be encouraged.

Horticulture is a source and sink of carbon. The long-time sequestration of carbon in fruit crop (perennial nature) blocks carbon adequately and releases lower amounts of GHGs than agricultural crops. Minimum tillage and low nitrogenous fertilizer application will reduce the chances of GHG emission during field interventions. It is important to adopt the Integrated Disease and Pest Management (IDPM) to protect horticultural crops in changing climate conditions. Seasonal forecasting is important to manage the horticultural sector. The heat- and drought-resistant seed and plant varieties need to be brought to the knowledge of farmers. Furthermore, the post-harvesting approach is also very important, as most vegetable crops are highly perishable. The cultivation of medicinal and aromatic plants in areas unsuitable for vegetable cultivation should also be focused with fruit trees in agri-horticultural format, which makes crops more resilient to adverse climatic situations.

Animal Husbandry

Climate change has a major impact on the livestock sector. The rise in temperature may lead to the spread of diseases, which are also a major source of CH_4 emission. Therefore, adaptation and mitigation are both important. To reduce CH_4 emission in the animal husbandry sector, the farmer should hold fewer productive animals than inferior animals. It has been observed that crossbred cattle are more productive but difficult to be acclimatized in unusual climatic conditions with limited facilities, yet require extensive care. Therefore, it would be better to promote crossbred cattle in areas having good connectivity and access to various facilities. Area-wise breeding is a strategy to be adopted along with guaranteed health coverage of livestock animals.

To reduce the emission of CH_4, animals should be protected from heat stress and invasion of new diseases. Encouragement of stall feeding, proper vaccinations, and disease investigation and treatment are important. To mitigate the effects of heat stress, continuing with traditional wooden and grass-made sheds and roofs for animals are better practices than constructing concrete structures. Furthermore, heat-resistant breeds of animals may be developed. Studies on genotype and environment interactions are needed to explore the appropriateness of breed for varying agro-climatic and sub-agro-climatic zones of the country.

In India, the local feed and fodder management has not yet received enough attention. It has been observed that the climatic conditions are limiting the existing resources of feed and fodder. In this context, it would be worth to analyse the local feed and fodder and current practices for a proper, balanced diet of livestock. It is necessary to protect the existing grazing fields through promotion and development of pasture land and following up with existing rules and developing fodder varieties as per the demand and local needs to protect the deterioration of forest area as well as pasture lands. Mixed farming is suitable both for crops and for domestic animals—one to protect the crop from the adverse climate and the other to provide variable and nutritive feed to the animals. Practices like pisciculture and poultry farming need to be promoted in this direction. Moreover, practising agri-silvipasture and utilizing wasteland are also better solutions for increasing fodder availability. For mitigating climate change in the livestock sector, the breed needs to be improved. The unsafe disposal of animal carcasses is a factor that enhances the emission of GHGs; hence, proper disposal of dead animals is required.

The management of dung and urine in dairy waste is an important aspect to maintain hygienic conditions. Dung and urine management should be taken care of by promoting biogas plants and household manufacture of vermicomposting fertilizer. The importance and valuable properties of cattle urine as potential insecticide/pesticide should be brought to the knowledge of the farmers. It has been noticed that the management of water is improper in most domestic farms, and this creates a suitable platform for water-borne diseases. Along with severe temperature and excess rainfall, it leads to damage in livestock. For this, management of waste water in dairy farms through suitable drainage systems is an environment-friendly option. Another option is to collect all the excess water from the livestock pens and either recycle/reuse or divert to soak pits, from which water is allowed to percolate down and the residue can be used for composting.

Fisheries

The impact of climate change in Indian coastal areas and the fisheries sector can be lethal. Due to rising temperature, it could force the fish community to migrate from one place to another. Global warming may directly influence inland fisheries by changing breeding behaviour and influencing the reproductive cycle of fish. An integrated aquatic ecosystem management approach is needed for coastal and inland fisheries to adapt to changing climate and minimize risk from natural hazards. There is also a need for a national climate change adaptation for the fisheries sector, with an inclusion of fisheries and aquaculture on the same platform. There is a gap in proper vigilant planning for the aquatic ecosystem and fisheries may suffer as a result of adaptation measures applied by other sectors such as increased use of dams and hydro power in catchments with high rainfall, or marine wind farms (FAO 2009). Herbivorous fish are suitable for cultivation to lower climate change impact. Moreover, they are more nutritious, having a low carbon footprint.

Excess fishing in a particular zone has a negative effect on the fish population. Therefore, the fishing spot should be demarcated and the desired number of fish may be collected from the unit area of a region. It is necessary to provide education at school level in climate change, for better understanding of the importance of fisheries and the climatic vagaries. Scientific and economic studies estimating the carbon sequestration potential of the aquatic ecosystem are important for handling climate change. The strategies of fish-rearing practices in an integrated manner, with management of river basins and watersheds are also an important approach. Fisherman should be given detailed knowledge on tools and techniques for sustainable harvesting of resources and getting fair benefits from the fisheries sector.

Water Resources

As water is the important life-support system for the agriculture sector, it is necessary to follow and develop specific tools and techniques for its conservation and management. Global warming has a direct impact on the depletion of the water table; it has dried existing water sources. In such circumstances, the adoption of rainwater harvesting technology, furrow and ridge irrigation systems and micro-level irrigation are important to consider. A rainwater harvesting tank can be constructed and dugout at village or individual level, and used to store rain water for future use. Despite the fact that in the Indian agricultural system, canal and underground (tube well) water is used for irrigation, efficient use of water still remains a major issue. It has also been seen that water-lifting devices, such as tube-well pumps, used to extract the ground water, are of poor quality and consume large amount of electricity. Also, the excess water taken from the ground sub-surface is not used efficiently. Therefore, judicious use of water in the agricultural field is important.

CONCLUSION

Although information has been generated and compiled on climate change in the last few decades, many questions remain unanswered. The tasks of mitigation and adaptation in Indian agriculture require widespread knowledge and awareness not only in the scientific and administrative community but also at the farmer's level. Countless effects are likely to be projected on the Indian agricultural system due to climatic vagaries. However, these effects are weakly understood and our projections on climate change are still not very comprehensive. The factual impact of climate change on Indian agriculture is difficult to predict and becomes more complex in different agro-climatic and agro-ecological regions. The facts cannot be ignored in the rapid changing climatic conditions of the past few decades, particularly the rise in temperature and abnormality in rainfall patterns.

Adaptation is the major corridor to pass through for the future sustenance of Indian agriculture. The extent of adaptation to crops and livestock depends on our homework to face the challenges of climate change by means of scientific, genetic, technical, and indigenous knowledge. On the other hand, mitigation is equally important, as agriculture and livestock are among the major sources of CH_4 emission. But, it is more important to develop environment-friendly agricultural interventions for a green and clean future. The implementation of adaptation and mitigation pathways by the Indian farming community needs attention from policy-makers and planners. Sustainable and resilient agriculture practices may feed farmers and every person in the country and eventually provides goods and services at national and global levels.

ACKNOWLEDGEMENTS

The author is thankful to the Director, Indian Institute of Forest Management, Bhopal, India, for encouragement and support. Constructive suggestions by an anonymous reviewer to improve the manuscript are duly acknowledged.

REFERENCES

Daw, T., W. N. Adger, K. Brown, and M. C. Badjeck. 2009. Climate change and capture fisheries: potential impacts, adaptation and mitigation. In *Climate Change Implications for Fisheries and Aquaculture Overview of Current Scientific Knowledge*, edited by K. Cochrane, C. De Young, D. Soto and T. Bharati, pp. 107–150. [FAO Fisheries and Aquaculture Technical Paper No. 530, Rome, Italy]

DAC (Department of Agriculture and Cooperation). 2010. *Annual Report*. New Delhi: DAC, Ministry of Agriculture, Government of India

FAO (Food and Agriculture Organization). 2009. *Fisheries and Aquaculture in our Changing Climate*. Rome: FAO. Details available at <ftp://ftp.fao.org/FI/brochure/climate_change/policy_brief.pdf>

IPCC (Intergovernmental Panel on Climate Change). 2007a. *Climate Change 2007: The Physical Science Basis. Contribution of Working Group I to the Fourth Assessment Report of the Intergovernmental Panel on Climate Change*, edited by S. Solomon, D. Qin, M. Manning, Z. Chen, M. Marquis, K. B. Averyt, M. Tignor, and H. L. Miller. Cambridge, UK, and New York, USA: Cambridge University Press

IPCC (Intergovernmental Panel on Climate Change). 2007b. *Climate Change 2007: Synthesis Report. Contribution of Working Groups I, II and III to the Fourth Assessment Report of the Intergovernmental Panel on Climate Change*, edited by R. K. Pachauri and A. Reisinger. Geneva, Switzerland: IPCC

INCCA (Indian Network for Climate Change Assessment). 2010a. Climate change and India: a 4 × 4 assessment. A sectoral and regional analysis for 2030. In *Indian Network for Climate Change Assessment Report.* New Delhi: Ministry of Environment and Forests, Government of India

INCCA (Indian Network for Climate Change Assessment). 2010b. India: greenhouse gas emission 2007. In *Indian Network for Climate Change Assessment Report.* New Delhi: Ministry of Environment and Forests, Government of India

Jepma, C. J., and M. Munasinghe. 1998. *Climate Change Policy: Facts, Issues and Analyses.* Cambridge, UK: Cambridge University Press

Joshi, A. K., and P. K. Joshi. 2011. A rapid inventory of indicators of climate change in the Middle Himalaya. *Current Science* 100(6):831–833

Mohanty, B. P., S. Mohanty, J. K. Sahoo, and A. P. Sharma. 2010. Climate change: impacts on fisheries and aquaculture. In *Climate Change and Variability*, edited by S. Simard, pp. 119–138. InTech-Open Access Publisher

Ravindranath, N. H., and J. Sathaye. 2002. *Climate Change and Developing Countries*. Dordrecht: Kluwer Academic Publisher

Sathaye, J., P. R. Shukla, and N. H. Ravindranath. 2006. Climate change, sustainable development and India: Global and national concerns. *Current Science* 90(3):314–325

Sharma, S. C., and R. K. Roy. 2010. Impact of global climate change on floriculture in India. *EnviroNews: Newsletter of International Society of Environmental Botanist, India* 16(1)

Sinha, M., D. K. De, and B. C. Jha. 1998. *The Ganga: Environment and Fishery*. Kolkata: Central Inland Fisheries Research Institute

Swaminathan, M. S. 2010. Safeguarding national food security in an era of climate change. In *Agriculture Year Book, 2010*, pp. 6–7. New Delhi, India

Vass, K. K., M. K. Das, P. K. Srivastava, and S. Dey. 2009. Assessing the impact of climate change on inland fisheries in River Ganga and its plains in India. *Aquatic Ecosystem Health & Management* 12(2):138–151

CHAPTER 19

Climate Change Mitigation: Potential of Growing Tree Species of Medicinal Value

Suresh Chauhan, Mohit Gera, and Ashish Aggarwal

INTRODUCTION

India is one of the world's 12 mega biodiversity countries (Joy, Thomas, Mathew, *et al.* 2001) with 17 000 species of higher plants, of which 7500 are known for their medicinal uses (Shiva 1996). This proportion of medicinal plants is the highest known for their medicinal properties in any country. Among these species, about one third each are trees and herbs, while the rest 20% are shrubs, 12% climbers, and 3% others. An estimated 70% of the medicinal plants are extracted using destructive harvesting, threatening the genetic diversity of the species as well as posing challenges to the sustainable supply of raw material. In the wake of the recent surge in demand for traditional medicines, financial gains have obtained precedence over the traditional-use pattern (Dhar, Manjkhola, Joshi, *et al.* 2002), pushing many species to the brink of extinction. Over 90% of the medicinal plants used by pharmaceutical industries are sourced from forests. Almost 800 species are industrially used, of which only about 20 species are commercially cultivated. It is understood that the need of medicinal plant parts cannot be met by forests alone, even with improved management and practices. Therefore, the potential to grow medicinal plants on private lands continues to exist.

Besides, in the management and cultivation of medicinal tree species, the estimation of the carbon sequestration potential is a neglected aspect. This is of particular interest in relation to the role the forestry sector can play to mitigate CO_2 emissions (IPCC 2007). More specifically, more than a third of the medicinal plant species are trees and, thus, have substantial potential of carbon sequestration. This aspect of medicinal plants adds a new dimension to the sustainable use of such resources, which can provide additional benefits in the form of recurring income to poor and marginal farmers through carbon revenues. Recent experiences in forestry clean development mechanism (CDM) project development at the global level have also shown that projects involving low-income communities and native species associated with recurring benefits are likely to be more successful than projects involving commercial tree species. This is because fast-growing commercial tree species are generally grown for shorter rotations, resulting in lower carbon sequestration potential. Also, they are often associated with high financial returns, which makes the project difficult to address the additionality aspect and, therefore, may not be accepted for registration (UNFCCC 2008). However, recent research findings have shown that tree species of medicinal value have higher levels of carbon sequestration due to their longer growth periods and are therefore well-suited to the CDM project framework (Gera and Chauhan 2010).

Despite modalities and procedures in place for the development of forestry sink projects under CDM, there is less focus on projects involving tree-based medicinal-plant-growing interventions. This may be due to the lack of information on various technical aspects, including carbon sequestration potential and likely carbon benefits, besides the silvicultural practices under different plantation interventions. Thus, tree species, including those with medicinal values, sequester a significant amount of carbon over decades or even centuries of their growth, and, therefore, demand the attention of the research community to investigate their carbon sequestration potential, cost effectiveness, and other methodological issues pertaining to CDM (TERI 2010). The findings presented in this chapter are mainly driven by this research gap of assessing the carbon sequestration potential of tree species of medicinal value as a climate change mitigation option.

STUDY AREA

The field study under this research was conducted in three states of Northern India—Punjab, Haryana, and Uttarakhand (Figure 1) between April and September 2008. In Uttarakhand, field surveys were carried out in the Udham Singh Nagar and Nainital districts of the Terai region, a relatively flat terrain in mountain-dominated topography. Kothkhara and Tanda forest sites were covered in Udham Singh Nagar district, while Kishanpur and Paulgarh forest sites were surveyed in Nainital district. In Haryana, the case study sites were taken from Solan KaTapar, Chaper, Mallah, Uttaroan, and Sehat forests of the hilly terrain of the Mourni Pinjore area in Panchkulla district and the plains of Gual Pahri in Gurgaon district. In Punjab, the hilly terrain of Dasuya Forest in Hoshiarpur district, and the plains of Sukhana Wildlife Century, Botanical Garden, and Rose Garden in Chandigarh were covered. The details of the forest sites surveyed in different states are provided in Figure 1. The research study in the selected locations focused on analysing the carbon sequestration potential of eight medicinal tree species under plantation on the forest land. The medicinal tree species studied for this research include *amla* (*Emblica officinalis*), arjun (*Terminalia arjuna*), *bael* (*Aegle marmelos*), *bahera* (*Terminalia belerica*), *harar* (*Terminalia chebula*), *jamun* (*Syzigium cuminii*), *neem* (*Azadirachta indica*), and *reetha* (*Sapindus mucorossi*). These species were selected after detailed discussion and consultation with the scientists and experts of National Medicinal Plant Board (NMPB) and field officials, because of the wide abundance of these species across the agro-climatic zones within the selected area and their good carbon storage potential capacity.

Figure 1 *Location of case study areas*

METHODOLOGY

Shortlisting Plantation Interventions and Field Observations

Field surveys in the selected states were organized in a phased manner. In Uttarakhand, it was organized during April and May 2008, followed by Haryana in June and July 2008, and finally in Punjab .during August and September 2008. It started with the collection of secondary data on the selected medicinal tree species from the respective state forest departments. The acquired data was subject to preliminary analysis to understand the distribution pattern of medicinal tree species according to different agro-climatic zones, plantation models, and age classes. This was followed by developing a detailed sheet for field data collection on estimating the eligible carbon pools: above ground biomass (AGB), below ground biomass (BGB), woody litter (WL), and soil carbon (SC), the costs involved in raising and maintaining the plantations, and the possible associated recurring benefits received from fruits, seeds, leaves, bark, and fuelwood during the rotation period. The data-recording requirements, as laid out in the modalities and procedures under CDM, were followed to assess the carbon pools. A minimum of

3 quadrats of size 25 m × 20 m for each species covering different age classes were taken to record observations. To estimate AGB, the stands of different age classes were selected and the data on the girth and height of randomly selected trees was recorded. The cost-effectiveness of the selected plantations was recorded with detailed discussions by developing a questionnaire and conducting interviews with the forest staff and local people (Table 1).

The Mean Annual Increment (MAI) of all the selected tree species was calculated using the local volume equations (FSI 1996). The calculated stem volume was then converted into dry biomass by multiplying it with biomass expansion factor and wood density as per the Good Practices Guidelines provided by the Intergovernmental Panel on Climate Change (IPCC 2006) to obtain AGB. While BGB was calculated using the IPCC default value (AGB × 0.27), carbon captured was calculated by multiplying the biomass by 0.45, again an IPCC default value. Besides, three soil samples were collected from each location where measurements were recorded for plantations, as well as for control from nearby areas with no plantations. For each location, samples were collected at three different depths of 15 cm, 30 cm, and 45 cm and mixed to form a composite sample. The collected samples were labelled and analysed for soil organic carbon by applying the method given by Walkley. Data on woody litter was estimated by laying out 1 m × 1 m quadrats within the selected tree quadrats, collecting the freshly fallen woody litter, and extrapolating the values in terms of tonnes per hectare.

Carbon Mitigation Potential and Cost-Effectiveness

The carbon mitigation potential and cost-benefit analysis of the selected plantation interventions were calculated using Excel™-sheet-based model, PRO-COMAP (Project Comprehensive Mitigation Assessment Process) developed by Sathaye, Makundi, and Andrasko (1995). Also, there is an attempt to consider the implications and other uses of this model application as a policy tool and its usefulness in practical implementation. This model has been designed specifically to undertake a comprehensive assessment of the role of tree plantations in climate change mitigation. It includes four of the five carbon pools defined by UNFCCC 2003; these are AGB,

Table 1 *Key parameters input in PRO-COMAP model on forest land based on species*

Parameter/Species	*Amla*	*Bael*	*Arjun*	*Bahera*	*Harar*	*Jamun*	*Neem*	*Reetha*
Uttarakhand								
Baseline soil carbon (tC/ha)	46.80	28.80	38.70	45.00	45.00	41.40	37.80	45.00
MAI for AGB (tB/ha/y)	01.43	02.00	06.04	06.79	05.15	04.20	02.15	05.15
Rotation Age (no. of years)	30.00	30.00	50.00	50.00	50.00	50.00	50.00	50.00
Soil carbon uptake (tC/ha/y)	0.10	0.05	0.16	0.24	0.30	0.30	0.36	0.30
Woody litter (t/ha/y)	0.30	0.12	0.27	0.21	0.20	0.11	0.20	0.20
Haryana								
Baseline soil carbon (tC/ha)	45.90	27.12	17.55	39.40	21.15	17.55	25.50	28.00
MAI for AGB (tB/ha/y)	02.00	03.50	07.48	04.91	06.00	05.30	04.16	05.00
Rotation Age (no. of years)	30.00	30.00	50.00	50.00	50.00	50.00	50.00	50.00
Soil carbon uptake (tC/ha/y)	0.12	0.16	0.14	0.20	0.25	0.21	0.20	0.19
Woody litter (t/ha/y)	0.25	0.09	0.10	0.10	0.07	0.19	0.15	0.07
Punjab								
Baseline soil carbon (tC/ha)	22.40	21.60	21.60	24.75	15.75	18.50	18.90	23.40
MAI for AGB (tB/ha/y)	02.32	02.85	06.10	04.11	05.12	03.25	04.20	03.50
Rotation Age (no. of years)	30.00	30.00	50.00	50.00	50.00	50.00	50.00	50.00
Soil carbon uptake (tC/ha/y)	0.66	0.45	0.79	0.53	0.43	0.26	0.45	0.38
Woody litter (t/ha/y)	0.23	0.28	0.32	0.27	0.21	0.18	0.26	0.26

BGB through an expansion factor, WL, and SC (Khatun, Valdes, Knorr, *et al.* 2010). Dead Organic Matter (DOM), which is the fifth eligible carbon pool, however, is not included in the present study. The model uses data on selected carbon pools as collected from the field: AGB, BGB, SC, and WL, along with data on costs and benefits throughout the rotation period. To estimate the future investment necessary for implementing plantation interventions and the effect of those investments on the plantation rate, the linear model establishes a baseline scenario with no financial revenues from carbon (Ravindranath and Murthy 2007).

All costs on planting and maintenance of each plantation intervention, including the cost for planting, silvicultural operations, thinning, harvesting, recurring costs on maintenance and protection, and monitoring biomass were considered. The benefits were calculated on the basis of prevailing market rates of tree products besides carbon benefits, which were estimated separately. There were local consultations held and there was also reference to literature to estimate the annual yield of fruits, seeds, and other products, as well as fuelwood on pruning. The cost-effectiveness of the selected plantation interventions was estimated using three different discount rates: 6%, 8%, and 10% for amla and bael, which are short-rotation crops (30 years), and 8%, 10%, and 12% for other species, which are long-rotation crops (50 years).

The PRO-COMAP model runs the analysis in two parts which are as follows.

1. The baseline scenario-carbon mitigation under the business-as-usual scenario, where no deliberate effort is made to reduce carbon emission.
2. The mitigation scenario-mitigation options such as tree plantation activities are undertaken for carbon mitigation over a defined period of time.

The analysis period selected for the present study was 2008–38 (30 years). The output of the model consists of the annual incremental carbon sequestered, values on cost-effectiveness indicators, and expected benefits under the selected carbon price scenarios for each plantation intervention. The relevant data provided as key inputs to the model are presented in Table 1.

The model estimates the carbon stock changes based on rotation age, mean annual increment of species, decomposition rates of litter, and accumulation time of carbon in the soil. It was assumed that the baseline carbon stock would be static for all plantation interventions, as they are to be taken up on forest lands devoid of any appreciable tree biomass. The static baselines are valid under CDM for projects where no changes in anthropogenic or natural activity are expected over the project period (Richards and Andersson 2001). Under the mitigation scenario, the major changes in C stock accrues due to the changes in AGB and BGB carbon stocks. The fallen woody litter from time to time also contributes to the carbon pool, which, though minimal, has been accounted for, along with soil uptake on an annual basis.

RESULTS

Annual Carbon Increment and Likely Monetary Benefits

In Uttarakhand, the maximum carbon sequestration potential was observed in the *bahera* plantation, followed by *arjun*, *reetha*, *harar*, *jamun*, *neem*, *bael*, and *amla* plantations with the potential to sequester carbon to the extent of 105, 92, 83, 83, 69, 41, 29, and 23 tC/ha, respectively. In Haryana, the maximum carbon sequestration potential was observed in *harar*, followed by *arjun*, *jamun*, *bahera*, *neem*, *bael*, *reetha*, and *amla* plantations with 123, 113, 112, 77, 66, 49, 43, and 30 tC/ha, respectively. In Punjab, the maximum carbon sequestration was observed in *arjun*, followed by *harar*, *neem*, *reetha*, *bahera*, *jamun*, *bael*, and *amla* plantations at 137, 86, 73, 61, 59, 54, 48, and 48 tC/ha, respectively. It is evident from the results that tree species such as *arjun*, *harar*, *bahera*, *neem*, and *jamun* have the capacity to store more carbon than the *amla* and *bael* species. This is obviously due to higher woody growth rates in these species compared to *amla* and *bael*.

The annual carbon sequestration potential of the selected plantation interventions was also estimated. It varies from *bahera* plantation with 3.50 tC/ha/y to *amla* with 0.77 tC/ha/y in Uttarakhand, from 2.46 tC/ha/y for *harar* to 0.96 tC/ha/y for *amla* in Haryana, and from 2.87 tC/ha/y for *arjun* to 0.96 tC/ha/y for *amla* in Punjab. Similarly, the annual sequestration of CO_2 was assessed by multiplying the carbon values by 3.67. In Uttarakhand, the annual incremental sequestration potential of CO_2 varies from 2.83 tCO_2/ha/y for *amla* plantation to 12.85 tCO_2/ha/y for *bahera* plantation, while in Haryana it varies from *harar* (9.03 tCO_2/ha/y) to *amla* (3.53 tCO_2/ha/y), and in Punjab from *amla* (3.53 tCO_2/ha/y) to *arjun* (10.53 tCO_2/ha/y). The likely monetary benefits of carbon sequestration were calculated under two price scenarios. The first was calculated assuming a rate of \$5 per tCO_2 and the second, at the rate of \$10 per tonne, assuming a dollar-rupee exchange rate of \$1 = ₹46. In Uttarakhand, the likely monetary benefits under the first scenario vary from *bahera* (₹2954) to *amla* (₹650), while in Haryana, it varies from *harar* (₹2076) to *reetha* (₹726) and in Punjab, from *arjun* (₹3858) to *amla* (₹1351). The price in each plantation intervention would double under the second price scenario (Figures 2, 3, and 4).

Internal Rate of Return With and Without Carbon Benefits

The project cost-effective indicator, the Internal Rate of Return (IRR) with and without carbon benefits was calculated separately for all the eight plantation interventions in the three states, on a per hectare basis.

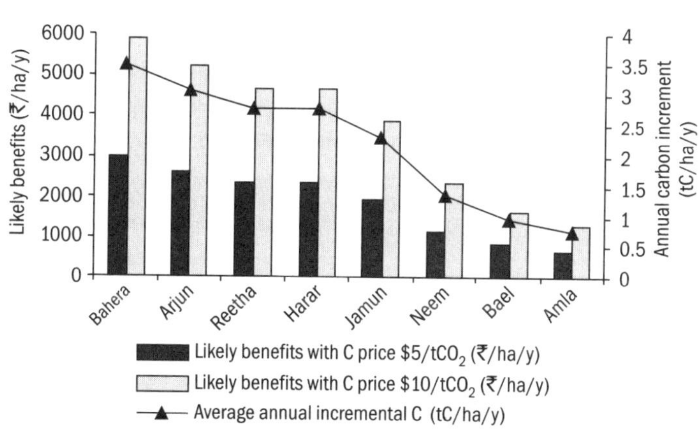

Figure 2 *Annual carbon increment and likely carbon benefits under two carbon price scenarios in Uttarakhand*

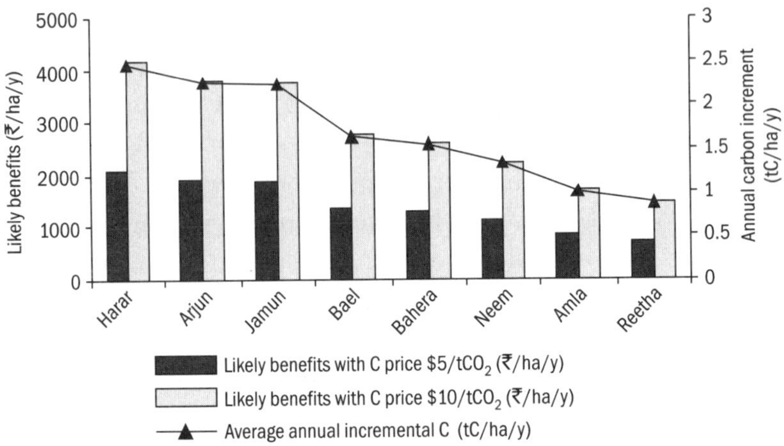

Figure 3 *Annual carbon increment and likely carbon benefits under two carbon price scenarios in Haryana*

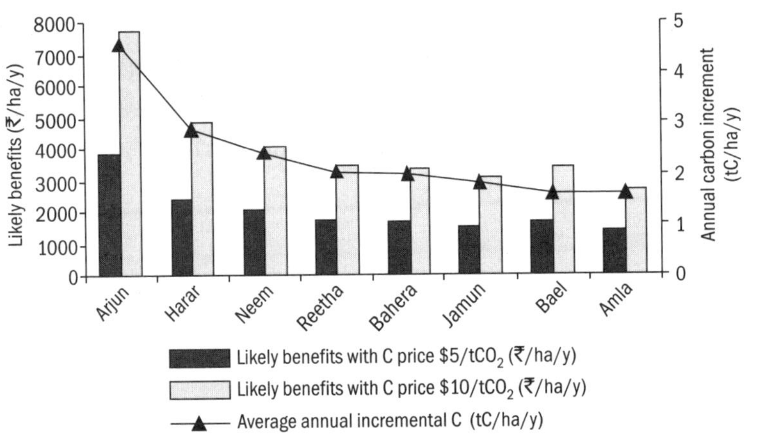

Figure 4 *Annual carbon increment and likely carbon benefits under two carbon price scenarios in Punjab*

In Uttarakhand, the IRR without carbon benefits was observed to be maximum in *bael* (52.90%) followed by *amla* (52.80%), *neem* (39.80%), *jamun* (38.80%), *harar* (37.30%), *reetha* (36.90%), *bahera* (35.50%), and the lowest in *arjun* (31.80%). With carbon benefits, the net increase in IRR for *bael, amla, neem, jamun, harar, reetha, bahera,* and *arjun* was recorded as 3.8%, 2.6%, 4.2%, 8.3%, 9.1%, 10.5%, 12.4%, and 11.9%, respectively.

In Haryana, the IRR without carbon benefits was the maximum in *amla* (43.70%) and *bael* (42.90%) and the lowest in *arjun* (29%). However, with the addition of carbon benefits, the IRR increased to a maximum in *harar*, *jamun*, and *arjun* by 6.3%, 6%, and 5.6%, respectively.

Similarly, in Punjab, the IRR without carbon benefits was found to be the maximum in *amla* (46.2%) and *bael* (44.9%) and the lowest in *arjun* (29.7%).With the addition of carbon benefits, the IRR increased to the maximum in *neem*, *harar*, and *arjun* by 7%, 7.3%, and 14.8%, respectively (Table 2).

NET PRESENT VALUE

Another cost-effective indicator, the net present value (NPV) was also calculated for all the interventions in the three states at three discount rates: 6%, 8%, and 10% for short-rotation crops (amla and bael) and 8%, 10%, and 12% for the remaining long-rotation crops. In Uttarakhand, the maximum NPV was recorded in *amla* (₹557 783 per ha) and *bael* (₹424 780) plantations at 6% discount rate, while the minimum was seen in *arjun* (₹89 872), *bahera* (₹128 495), and *reetha* (₹129 959) at 12% discount rate.

In Haryana, the maximum NPV was recorded for *amla* (₹517 251) and *bael* (₹411 409) at 6% discount, and the least in *arjun* (₹86758) and *bahera* (₹133 738) at 12% discount. Similarly, in Punjab, *amla* (₹520 773) and *bael* (₹413 296) provide the maximum returns at 6% discount, while *arjun* (₹88 114) and *bahera* (₹135 524) provide the lowest at 12% discount (Table 3).

DISCUSSION AND CONCLUSION

The present study indicates that *arjun*, *bahera*, *harar*, *jamun*, and *neem* plantations have a higher carbon sequestration potential than other species of medicinal importance, while *amla* and *bael* plantations were observed to be financially more attractive and can provide reasonable positive returns even without carbon benefits. *amla* and *bael* are short-rotation, slow-growing crops and, thus store Less carbon than the other long-rotation tree species. *Arjun*, *bahera*,

Table 2 *Potential carbon benefits and increase in IRR on forest land*

	Uttarakhand			Haryana			Punjab		
Plantation intervention	*IRR without carbon benefit (%)*	*Likely benefit with C price \$5/tCO_2 (₹/ha/y)*	*IRR with carbon benefit (%)*	*IRR without carbon benefit (%)*	*Likely benefit with C price \$5/tCO_2 (₹/ha/y)*	*IRR with carbon benefit (%)*	*IRR without carbon benefit (%)*	*Likely benefit with C price \$5/tCO_2 (₹/ha/y)*	*IRR with carbon benefit (%)*
Amla	52.80	650.00	55.40	43.70	844.00	46.00	46.20	1351.00	50.70
Bael	52.90	819.00	56.70	42.90	1384.00	47.60	44.90	1688.00	51.60
Arjun	31.80	2591.00	43.70	29.00	1907.00	34.60	29.70	3858.00	44.50
Bahera	35.50	2954.00	47.90	33.40	1300.00	37.00	34.70	1663.00	40.10
Harar	37.30	2338.00	46.40	34.70	2076.00	41.00	36.10	2423.00	43.40
Jamun	38.80	1941.00	47.10	35.50	1891.00	41.50	37.00	1519.00	42.40
Neem	39.80	1156.00	44.00	36.70	1114.00	39.70	38.30	2051.00	45.30
Reetha	36.90	2338.00	47.40	40.80	726.00	43.00	42.50	1714.00	48.80

Table 3 *Net present value in different discount rates on forest land*

Intervention	*Discount rates (%)*	*Net present value (₹/ha)*		
		Uttarakhand	*Haryana*	*Punjab*
Amla	6	557 783.00	517 251.00	520 773.00
	8	424 523.00	390 388.00	393 773.00
	10	326 816.00	297 601.00	300 860.00
Bael	6	424 780.00	411 409.00	413 296.00
	8	328 337.00	312 019.00	313 871.00
	10	256 412.00	238 777.00	240 596.00
Arjun	8	166 324.00	162 771.00	164 305.00
	10	122 053.00	118 770.00	120 211.00
	12	89 872.00	86 758.00	88 114.00
Bahera	8	231 259.00	245 309.00	247 161.00
	10	171 874.00	180 632.00	182 450.00
	12	128 495.00	133 738.00	135 524.00
Harar	8	262 253.00	257 157.00	259 009.00
	10	195 670.00	190 917.00	192 736.00
	12	146 952.00	142 544.00	144 329.00
Jamun	8	263 771.00	258 196.00	260 048.00
	10	197 161.00	191 937.00	193 756.00
	12	148 417.00	143 545.00	145 331.00
Neem	8	265 502.00	264 783.00	266 635.00
	10	199 607.00	198 728.00	200 546.00
	12	151 193.00	150 036.00	151 822.00
Reetha	8	232 777.00	369 720.00	371 572.00
	10	173 364.00	277 691.00	279 509.00
	12	129 959.00	210 251.00	212 037.00

harar, and *jamun* are comparatively fast growing, with longer rotation periods and can, therefore, store more carbon for longer periods. The cost-benefit analysis indicated that even without carbon benefits, the maximum returns may be gained by the growing tree species of *amla* and *bael*, compared to the other long-rotation species.

The IRR is an indicator of the threshold interest rate below which the loan for investment can be obtained to keep the project financially viable. The IRR without carbon benefits for *amla* varies from 43.70% to 52.80% and for *bael* from 42.90% to 52.90%. With the addition of carbon benefits, it increased from 46% to 55.40% and 47.60% to 56.70% for *amla* and *bael* plantations, respectively. The higher IRR rate without carbon benefits for these two species is because the fruits of these two species have very good market demand as well as price in North India. Despite the comparative cost of planting, maintenance, and other silvicultural activities, the higher returns in terms of fruit yield and price compared to the other species make these two interventions financially attractive even without carbon benefits. Also, grafted *amla* plants start yielding at 3–4 years and bael yields at 6–7 years of plantation, which is quite short compared to the rest of the species such as *harar*, *bahera*, *arjun*, and *jamun*, which start fruiting only after 8–10 years (Chauhan and Gera 2010).

The findings of this study are on similar lines to the studies conducted for different agro-forestry plantation models by Gera, Mohan, Bisht, *et al.* (2006), which indicated that the sequestration potential for poplar and eucalyptus species in Northern India varied from 1.42 tC/ha/y to 2.54 tC/ha/y with associated benefits from ₹1094 ha/y to ₹2198 ha/y. Another study conducted by Hooda, Gera, Vasistha, *et al.* (2005) also indicated that orchard planting such as mango, litchi, and citrus fruits can sequester carbon in the range of 0.20 tC/ha/y to 1.70 tC/ha/y with associated benefits of ₹154–₹1310 ha/y. The same study also reported that the carbon sequestration potential of agro-forestry species, such as poplar, eucalyptus, and teak was in the range of 1.42 tC/ha/y to 2.85 tC/ha/y with associated benefits in the range of ₹1094 ha/y to ₹2198 ha/y. The results obtained from the present study are comparable with these studies but one must keep in mind that compared to commercial tree species such as poplar and eucalyptus, medicinal tree species have more resilience towards price fluctuations owing to their long rotations, and are likely to face the least price volatility or sudden variation in demand/supply. In addition, growing medicinal tree species also has the advantage of providing sustainable livelihoods, biodiversity conservation, and filling the gap of demand and supply for raw material to pharmaceutical companies on a sustainable basis.

This is supported by the annual demand growth of *amla*, which was recorded as 22.5% between 2001/02 and 2004/05 (NMPB 2008). Also, the demand supply gap for other species such as *neem*, *bael*, and *harar* during 2001/02 was recorded as 1407, 901, and 510 tonnes, which increased to 13864, 2605, and 1365 tonnes, respectively, during 2004/05 within the country (CERPA 2001). Similarly, there is a huge demand for other raw material and processed products from many other tree species of medicinal importance.

The findings of the research study reveal that planting tree species of medicinal value offers a profitable source of carbon sequestration and can be a good option for mandatory markets like CDM or voluntary markets such as the Chicago Climate Exchange (CCX). The study has further revealed that the carbon sequestration levels not only depend on the growth, but also on the crop rotation and plant density. Apart from the recurring benefits of fruit and other medicinal parts, an increase in financial returns for providing ecosystem services through carbon sequestration can certainly motivate growers and stakeholders interested in undertaking the plantation of these tree species. The study has also demonstrated that planting trees of medicinal importance can result in the realization of appreciable carbon benefits on a recurring basis due to the large carbon sequestration capabilities of these plantations. The study has further shown that the recurring benefits from fruits, seeds, and other products associated with these plantation interventions, along with the additional returns from carbon benefits would motivate farmers to opt for plantation of tree species of medicinal value.

ACKNOWLEDGEMENTS

The article is based on the project 'Feasibility Study on Carbon Sequestration Benefits from Tree Species of Medicinal Value" funded by the National Medicinal Plants Board, New Delhi, Government of India. The authors acknowledge the support provided by the State Forest Departments of Uttarakhand, Haryana, and Punjab during the field study. They acknowledge C S Silori for reviewing the article and providing suitable comments.

REFERENCES

CERPA (Centre for Research Planning and Action). 2001. *Demand Study for Selected Medicinal Plants,* vol. I. New Delhi, India: CERPA

Chauhan, S., and M. Gera. 2010. Carbon benefits from growing trees of medicinal importance in Punjab. *Indian Forester* 136(7):861–876

Dhar, U., S. Manjkhola, M. Joshi, A. Bhatt, A. K. Bisht, and M. Joshi. 2002. Current status and future strategy for development of medicinal plants sector in Uttaranchal, India. *Current Science* 83(8):956–964

FSI (Forest Survey of India). 1996. *Volume Equations for Forests of India, Nepal and Bhutan.* Dehradun: FSI, Ministry of Environment and Forests, Government of India

Gera, M., and S. Chauhan. 2010. Opportunities for carbon sequestration benefits from growing trees of medicinal importance on farmlands of Haryana. *Indian Forester* 136(3): 287–300

Gera, M., G. Mohan, N. S. Bisht, and N. Gera. 2006. Carbon sequestration potential under agroforestry in Rupnagar district of Punjab. *Indian Forester* 132(5):543–555

Hooda, N., M. Gera, H. B. Vasistha, M. Chandran, and S. S. Rassaily. 2005. *Methodologies in Forestry Mitigation Project: Case Studies from Nainital and Udham Singh Nagar Districts of Uttaranchal.* Dehradun: Indian Council of Forestry Research and Education

IPCC (Intergovernmental Panel on Climate Change). 2007. *Climate Change 2007: Synthesis Report. Contribution of Working Groups I, II and III to the Fourth Assessment Report of the Intergovernmental Panel on Climate Change.* Edited by R. K. Pachauri and A. Reisinger. Geneva, Switzerland: IPCC

IPCC (Intergovernmental Panel on Climate Change). 2006. IPCC Guidelines for National Greenhouse Gas Inventories. [Prepared by the National Greenhouse Gas Inventories Programme, IPCC, IPCC/IGES, Hayama, Japan]

Joy, P. P., J. Thomas, S. Mathew, and B. P. Skaria. 2001. *Medicinal Plants: Tropical Horticulture,* vol. 2, pp. 449–632. Kolkata: NayaPrakash

Khatun, K., P. J. Valdes, W. Knorr, and R. K. Chaturvedi. 2010. Assessing the mitigation potential of forestry activities in a changing climate: a case study for Karnataka. *Forest Policy and Economics* 12:277–286

Ravindranath, N. H., and I. K. Murthy. 2007. Methodological issues in forestry mitigation projects: a case study of Kolar District. *Mitigation and Adaptation Strategies for Global Change* 12(6):1077–1098

Richards, K., and K. Andersson. 2001. The leaky sink: persistent obstacles to a forest carbon sequestration program based on individual projects. *Climate Policy* 1(1):41–54

Sathaye, J. A., M. Makundi, and K. Andrasko. 1995. A comprehensive mitigation analysis process (COMAP) for the evaluation of forestry mitigation options. *Biomass and Bioenergy* 8(5):345–356

Shiva, M. P. 1996. *Inventory of Forestry Resources for Sustainable Management and Biodiversity Conservation.* New Delhi: Indus Publishing Company

TERI (The Energy and Resources Institute). 2010. Project report on feasibility study on carbon sequestration benefits from growing tree species of medicinal value. New Delhi: TERI [Project Report No. 2007SF08, submitted to National Medicinal Plant Board]

UNFCCC (United Nations Framework Convention on Climate Change). 2003. Caring for climate: A guide to the Climate Change Convention and the Kyoto Protocol. [Issued by the Climate Change Secretariat, Bonn, Germany]

UNFCCC (United Nations Framework Convention on Climate Change). 2008. A/R methodological tool to identify the baseline scenario and demonstrate additionality in A/R CDM project activities. [EB 35 Report, Annex 19]

Walkley, A. 1947. An examination of methods for determining organic carbon and nitrogen in soils. *Journal of Agriculture Science* 25:298–609

Index

A

Above ground biomass (AGB), 302, 303, 304*t*, 305, 306

Aeromycoflora survey, 4, 6, 8

Agricultural sector in India, 283, 284, 289

  impact of climate change on, 272, 274, 278, 279

Agri-silvipasture, 293

Airborne fungi, 4, 6

Animal husbandry, 26, 183, 186, 283, 288, 289, 292-294

  and climatic change in India, 288

Arthropod diversity, 13

Artificial propagation, 99

Asia Fellowship programme, 150

Aspergillus, 5-8, 10

Ayurveda, 43, 165, 167

Ayurvedic medicine, 40, 43, 248

  demand for, 168

Ayurvedic preparation, 253

B

Balam, 251-252

  preparation of, 251

Ban Mae Sa Mai (BMSM), 150, 151, 152, 155*t*, 159

  agricultural practices in, 152

  conservation awareness in, 158-159

  effects of forest destruction on, 154

  household income in, 153, 158, 160

  institutional linkages of, 159

  major occupations in, 153

Below ground biomass (BGB), 302, 303, 305, 306

  default value of, 303

Bentham and Hooker's system of classification, 230

Berberis aristata, 29, 221, 230, 253

Bhaishaj Sangh, 169, 173

Bhotiyas, 212, 214, 216, 217, 219, 220, 224, 251-253

  semi-fermented food of, 252

  timur collection and its barter by, 212, 214, 220, 224

  uses of timur by, 212, 213, 215, 216-218

Bhyundar Valley, 13-22

see also Spiders

  spider fauna of, 17, 19, 21

altitudinal relationship between, 19, 21*f*
checklist of, 18*t*-19*t*
composition of, 17-21
density of, 21*f*, 22
family composition of, 20*f*
sample collection for the study on, 15
statistical analysis of, 15-16
location of, 15, 16*f*

Binsar Wildlife Sanctuary (BWS), 25-33
diversity dominance relationship in, 29
land use patterns in, 33
natality and mortality of seedlings and saplings in, 30,31*t*
wildlife species in, 27-28
woody vegetation in, 28-29
distribution and structure of, 28-29
species diversity and dominance in, 29, 30*t*, 32, 33

Biodiversity, 51
2010 target for, 115
and adaptation to climate change, 275
effect on ecosystem's reliability of, 265
effects of climate change on, 260-262
global estimates for, 52-54
impact of urbanization on, 113
in-situ conservation, 9, 99-100
levels of, 51, 52*f*
loss of, 52, 54

Biodiversity assessment, 52, 53*f*
components of, 53*f*

Biodiversity Characterization at Landscape Level (BCLL), 55, 59
geospatial model for, 59*f*

Biodiversity conservation, 8, 115, 118, 119, 120*t*, 130, 174, 196, 197,198, 245-254, 259-267, 322, 312
challenges of, 259-267
and climate change issues, 259-267
in community forest areas, 278
of fungal species, 9, 10
in Garhwal Himalayan region, 174
for local communities, 271, 273-278
and poverty alleviation, 130
in Phutoei National Park, 195-208
within tropical ecosystems, 263
in urban ecosystem, 113-115, 117, 133
in Uttarakhand, 283-293

Brahmi (*Centella asiatica)*, 249

C

Carbon mitigation, 303-305
and cost-benefit analysis, 303, 311

Carbon pools, 302, 303, 305
five types of, 303

Carbon sequestration, 294, 300, 301, 306, 307, 309, 312, 313

of agro-forestry species, 312
estimation of, 300
monetary benefits of, 306-307
in medicinal plants, 299, 300
net present value (NPV) of, 309, 311*t*
Cardamom-based agro-forestry, 277
Chamoli, 15, 173, 212, 213, 224, 247*f*
vegetation of, 213
Charak-ka-Daanda, 166
Chhattisgarh, 3, 8, 9, 37, 38, 39, 46, 55, 65
fungal diversity of, 4-8
Chicago Climate Exchange (CCX), 313
City Biodiversity Summit, 117
Clarkia pulchella, 93
commercial aspects of, 266
Clean Development Mechanism (CDM), 265-267, 300, 301, 302, 306
in Dongargarh, 3-5
Climate change,
adaptation and mitigation, 265-267, 273, 283-296
for agriculture, 283-296
for animal husbandry, 288, 292
for fisheries, 288, 289, 294
for horticulture, 286-287
for water resources, 295
defined, 289
effects on, 260-262
agrobiodiversity, 262
Arctic ecosystem, 262
fisheries, 294
floriculture sector of India, 287
Gaderi or Arabi, 286
Himalayan oak forests, 261
horticultural crops, 291-292
Indian agriculture, 286
photosynthesis, 261
species distribution, 261
in India, 284-285
indigenous coping strategies to, 275-278
and landscape changes, 133
negative impact on forest ecosystems, 272, 274
two main processes of, 289
Community forest management (CFM), 160, 161, 184, 201, 206, 275
see also Huai Hin Dam Village
and indigenous knowledge, 296
success of, 161
Community-initiated forest conservation, 149
Conference of Parties (COP-6), 245
Conservation Assessment and Management Plan (CAMP), 249
Convention on Biological Diversity (CBD), 117, 245

D

Dactylorhiza hatagirea, 249
Daru, 252
preparation of, 252

Deciduous forest in Chhattisgarh, 38
medicinal plants in, 37-48
distribution pattern of, 43*f*, 44-46,
endangered species of, 46
overexploitation of, 37, 43, 44, 47, 48
population of, 39-40
rarity and management of, 46-48
social dimensions of, 40-44
species used, 41*t*-42*t*
survey of, 39, 44
and problem of snakebites in, 43
Deomali hill range, 66
species found in, 66
Dodo, 260
Doi Suthep-Pui National Park (DSNP), 149-161
community-initiated forest conservation in, 149, 151
deforestation in, 149, 150
environmental services in, 157
field data collection from, 151-152
floral diversity in, 150
forest quality in, 154-156
forest restoration process in, 158,
income pattern in, 154*f*, 155*t*, 158
Dongargarh, 3, 4, 5, 7, 10 *see also* Fungi
fungal biodiversity of, 4-10, 6*t*
Dung and urine management, 293

E

ESTIMATE S version 8.0, 16
Ethnobotany, 227, 228
Ethno-taxonomy, 228
Eucalyptus plantation, 267, 312
Eutrophication, 96*f*
Evolutionary potential loss, 91, 95-97, 99
Excel™-sheet based model, 303

F

Feed and fodder management, 293
Fermented food and beverages, 251-252
traditional knowledge on, 251-252
Forest dependent communities, 223, 271, 272-273, 278
in Asia Pacific, 271
impact of climate change on, 272, 274, 278, 279
Forestry sink projects, 300
FORRU-CMU, 151
Fungi, *see also* Dongargarh
biodiversity, 4-10
in Dongargarh, 3-10, 6*t*
conservation of, 8, 9
role in ecology, 8
species diversity in India, 5-8

G

Gandhamardan hill range, 65
taxa of, 65
Genetic diversity, 47, 93-101, 299
conservation of, 99-101
implications of, 99-101
loss of, 93-97

low levels of, 97
Geographic information system tools, 54, 114, 119, 134*t*
Global warming, 294, 295
Green Data Book, 166
Greenhouse gas (GHG), 265-266, 284, 285, 290, 292, 293
emissions in India, 285, 290-293, 296

H

Habitat,
degradation, 133
loss, 8, 54, 93, 133
Habitat fragmentation, 91, 94, 99, 133, 137
affect on genetic variation, 94
Himalayan glaciers, 288
Hmong tribe, 149, 150, 151, 152, 154
history of, 152
HNB Garhwal University, 229
Homomorphic self-incompatibility, 96
Huai Hin Dam Village, 197, 198, 199*f*, 201
community forestry in, 201, 206
income patterns of, 197, 199, 203-205
JoMPA study in, 195-197, 205-208 *see also* Joint Management in Protected
Areas
Hutchinson's concept, 230
Hypochaeris salzmanniana, 97

I

Inbreeding depression, 92-95, 96, 98, 100 *see also* Plant extinction and vulnerability
effect of, 92-93
and extinction, 93-95
INCCA's assessment for India, 284, 285
India Himalayan Region (IHR), 227, 228
plant species in, 227, 229, 230, 231*t*, 240
protected area network in, 228
taxonomical studies in, 228
Indian climate change, 284-285
projections for 2030, 284
Indian Network for Climate Change Assessment (INCCA), 284
Indian Remote Sensing (IRS), 57
Indian System of Medicine (ISM), 167, 230
Integrated Disease and Pest Management (IDPM), 292
Intellectual Property Rights (IPR), 240
Intergovernmental Panel on Climate Change (IPCC), 272, 303
Fourth Assessment Report (AR-4), 271
Internal Rate of Return (IRR), 307-309, 310*t*, 311, 312
with and without carbon benefits, 300, 305, 309, 310*t*, 311, 312

IRS P6 LISS III, 57, 58*f*
IUCN, 46, 166, 253
IUCN Red List categories, 46, 166

J

Jaan, 251, 252
  cereals used for, 252
Joint Management in Protected Areas (JoMPA), 195-208 *see also* Phutoei National Park, and Thailand
  aim of, 195
  challenges to, 207, 208
  indicators used for, 197, 198, 202
  success of, 207-208
  in Thailand, 195-208

K

Kolmorogov–Smirnov test, 16
Kondakamberu hill range, 67
Kuth cultivation, 165-181
  activities conducted for, 175-178
  awareness camps for, 178-180
  cost-benefit analysis of, 175
  preparation of nursery for, 175, 177
  project area for, 175, 178
  seed germination for, 176, 177, 178*f*
  success story of, 174-175
  in *Uttarkashi*, 179*f*
Kyoto Protocol, 265, 266

L

Landscape ecological principle, 54

M

Mahendragiri hill range, 66
  species found in, 66
Mauritian calvaria, 260
Mean Annual Increment (MAI), 303, 306
Medicinal and Aromatic Plants (MAPs), 165, 167-173
  collection practices of, 169-170
  cultivation of, 170-173
    factors preventing, 171
  demand and prices of, 172
  endangered species of, 249
  exploitation of, 166, 168, 169
  ex-situ cultivation of, 174, 179
  and on-farm research issue, 171
  notable species of, 166
  in Uttarakhand, 166, 167-169, 170-173, 175, 179*f*,
Medicinal herbs, 249
Methane emission, 285, 288, 291, 292, 296
Millennium Ecosystem Assessment 2006, 129
Mixed population translocation, 100
Mulching techniques, 290
Munsyari, 213, 216, 219, 221, 224
  social groups residing in, 224

N

Nainital, 32, 247*f*, 301
Nanda Devi Biosphere Reserve (NDBR), 13, 14, 15, 227-241
  collection of plants from, 229
  ethno-taxonomical studies on, 227-241
  Folklore Survey in, 229
  medicinal plants of, 227-241

uses of, 230, 231*t*-239*t*

Nardostachys grandiflora, 249

National Centre of Fungal Taxonomy, 4

National Medicinal Plant Board (NMPB), 301

Natural Resource Property Right (NRPR), 240

Non-timber forest products, 154, 198, 223, 283

field surveys on, 246

study area for, 246

location of, 247*f*

O

Odisha,

biodiversity characterization of, 54, 55, 59

biological richness (BR) map of, 63, 64*f*

endemic/threatened species in, 52

total importance value (TIV) for, 61

false colour composite (FCC) image of, 57, 58*f*

landscape mapping in, 61-63

location map of, 56*f*

mapping of plant in, 51-67

medicinal plants richness areas, 65-67

new recorded floras in, 65

phytosociological analysis in, 59-61

Sal mixed deciduous forests in, 60, 61, 63

study area of, 55-56

use of remote sensing and GIS, 51-67

vegetation and land use map of, 62*f*

Open source/non-commercial GIS products, 134*t*-136*t*

Organic farming, 206, 290, 291

Outbreeding depression, 91, 92, 100

P

Parthenium hysterophorus, 71-88

adverse effects of, 72

allelopathic effects of, 72, 73, 88

common names of, 71

effect of aqueous extracts on, 72, 73, 74*f*

Brassica rapa, 72, 74*f*, 79-82, 80*t*

Lactuca sativa, 72, 73-76, 74*f*, 75*t*

Pisum sativum, 72, 74*f*, 82-85, 83*t*

Spinacea oleracea, 72, 74*f*, 77-79, 77*t*

material collection for the study on, 72

preparation of aqueous extracts from, 72-73

statistical analysis of, 73

Patharkot Village, 183-192, 188*f*, 189*t*,190*t*, 191*f*

agriculture crops in, 190

input and output ratio of, 190, 190*t*

area of, 184-185

cropping system of, 185-186, 186*f*

fodder demand and supply status in, 190*t*
fodder, 189
fuel consumption and supply in, 188
fuelwood species in, 189
horticulture production in, 187
household survey in, 185
income groups in, 189-190
livestock holding in, 186-187, 187*f*
main source of fuelwood, 188-189

Penicillium, 5, 6, 8, 10

Phutoei National Park, 195-208 *see also* Joint Management in Protected Areas (JoMPA), and Thailand
conservation status in, 202*f*, 207
community forestry initiatives in, 206
impact of, 197
improvement in forest quality of, 200-203
improvement in income pattern of, 203-205
natural dyes making plants in, 198, 203, 204*t*
reclassification of forest area of, 201
sale of NTFPs in, 203
village-level institutions in, 206

Phytodiversity, 55, 166
man-made threats to, 166

Pithoragarh, 220, 212, 213, 224

Plant extinction and vulnerability, 91-101 *see also* Inbreeding depression
effect of inbreeding on, 92-93, 94, 95, 96, 98, 100
genetic factors for, 91, 94, 95
and loss of,
evolutionary potential, 91, 95-97, 99, 101
genetic diversity, 95-96, 97, 98, 99, 100
self-incompatibility, 96-97

Polygonatum verticillatum, 249

PRO-COMAP model, 303, 304*t*, 305
two parts analysis by, 305
and their management, 195, 196, 206, 208
in Northern Thailand, 149-159
global network of, 263

Protected areas, 54, 149, 168, 195, 196, 206, 208, 253, 259, 263-265
global network of, 263
and their management, 195, 197, 206, 208
in Northern Thailand, 148, 150, 161

R

Red Data Book of Indian plants, 61, 166

Remote sensing, 51-67, 114, 120*t*-125*t*, 130-131, 132*t*, 134*t*,
application of, 118, 130-131
in urban studies, 131, 132*t*

Rheum australe, 240, 253
Rheum moorcrotianum, 253
Rights and resources initiative, 274
River Ganges, 288, 289

S

Sacred groves, 47, 276
Saptasajya and Kapilasa hill ranges, 67
Second opium war, 152
Seed bank genetics, 95
  research on, 95
Self-incompatibility, 96-97
loss of S-allele at, 96-97
Sez, 252
  extraction of, 252
Shannon–Wiener index, 16, 27, 60
Similipahar hill range, 65
  vegetation of, 65
Society for Himalayan Environmental Research (SHER), 174-175, 177, 180
  role of, 174
Soil carbon (SC), 302, 304*t*, 305
Spatial landscape analysis modelling (SPLAM), 59*f*
Species extinction, 91, 99 *see also* Plant extinction
Spiders, 13-22 *see also* Bhyundar Valley
  abundance of, 17, 20*f*
  diversity and distribution of, 13-14
  role in ecosystem, 13-14
  species in India, 14
SPSS version 16.0, 15
Subsistence-based agroecosystems, 184
System of Rice Intensification (SRI), 291

T

Thailand, 152, 195-208 *see also* Joint Management in Protected Areas (JoMPA), and Phutoei National Park
  Joint Management in Protected Areas (JoMPA) in, 195-208
  opium cultivation in, 152, 154
  self-initiated community conservation in, 149, 151
    see also Doi Suthep-Pui National Park
Timur (*Zanthoxylum armatum* DC), 211-224
  cash flow from, 221
  collection of, 211, 212, 214, 215, 216, 218-221
  conservation and management of, 221-222
  distribution pattern of, 215-216
  harvesting practice of, 218, 221, 222
  and management of, 218-219, 221
  plant association of, 221
  potential of, 221, 224
  social and economic dimension of, 219-221
  study on, 211-224
    field surveys for, 214
    results of, 215-216
    study area for, 213-214

uses of, 212, 213, 216-218

Tons watershed, 165, 174, 175, 178, 180

Traditional agroecosystems, 183-192

of Central Himalaya, 184, 190

Traditional agro-forestry models, 276

in Sri Lanka, 276

in eastern Himalayan countries, 277

Tree species of medicinal value, 299-313

carbon sequestration potential of, 300, 301, 306, 307, 309, 313

climate change mitigation potential of, 299-313

field surveys for, 301, 302

species selected for, 301

study on, 305, 309, 312, 313

area of, 301-302, 302*f*

U

Udham Singh Nagar, 247*f*, 301

UNFCCC 2003, 303

UN-Habitat, 148

United Nations Development Programme (UNDP), 272

Human Development Report of, 272

United Nations Environment Programme, 52

United Nations Framework Convention on Climate Change (UNFCCC), 265

Upland project, 159

Urban biodiversity, 114, 115-133, 120*t*-128*t*

considerations for, 117

management, 118, 120*t*-128*t*, 130-131

important issues for, 133

RS-GIS applications in, 130-131

overview of, 119, 120-129

recent studies related to, 120*t*-128*t*

and sustainability, 119, 129, 130

key attributes for, 129

use of geospatial technologies for, 119

Urban ecosystems, 113, 115, 117, 133

implications on environment, 116

use of geospatial tools for, 113-138

V

Vaidyas system of medicine, 248

collection of plants for, 251

species used as a tonic in, 249

study on, 246

Valley of Flowers National Park, 15, 228

Van Panchayats in Uttarakhand, 275

Vegetation in Indian sub-continent, 53

W

Widespread species, 97-98
  and rare species, comparison between, 97-98
  and rare species, genetic variation between, 98*f*
Woody litter (WL), 302, 303, 304*t*, 306
Wool dyeing, 253
  indigenous knowledge on, 253
  plants used for, 253
World Conservation Monitoring Centre (WCMC), 52

About the Editors

Dr Chandra Prakash Kala presently works with the Indian Institute of Forest Management, Bhopal, in the Ecosystem and Environment Management division. He obtained his PhD in Forestry from the Forest Research Institute, Dehradun and MSc in Botany from the H N B Garhwal University, Uttarakhand. Some of his earlier associations were with the Wildlife Institute of India, the G B Pant Institute of Himalayan Environment and Development, and the National Medicinal Plants Board, Ministry of Health and Family Welfare, Government of India.

Dr Kala has widely surveyed the various biogeographical areas, and has published over 150 research papers and articles in various national and international journals/magazines. He has also authored/co-authored five books that focus on biodiversity conservation, ecosystem and environment management, and sustainable development of medicinal and aromatic plants. He is in the advisory and editorial boards of many reputed national and international journals, including International Journal of Forestry Research, American Journal of Plant Sciences, International Journal of Ecology, and Journal of Biodiversity. He is the founder Executive Editor of the international journal, Medicinal Plants. He was awarded the prestigious 'ICFRE Award for Excellence' for his outstanding contribution in the field of forest conservation (biodiversity and ecology).

Dr Chandra Shekhar Silori presently works with RECOFTC–The Center for People and Forests, Bangkok. He obtained his PhD from the Saurashtra University, Gujarat, researching on people's dependence on forest resources of Mudumalai Wildlife Sanctuary, Tamil Nadu, and management of these resources to help balance the needs of local communities and subsequent conservation of elephant migratory routes passing through the sanctuary.

Dr Silori has vast experience of working in different ecosystems of India, Nepal, and South East Asia. While he was associated with the Wildlife Institute of India, he worked in two major biodiversity hotspots, the Western Ghats and the Himalayan mountains. He also worked in the arid and semi-arid regions of western India and mangrove ecosystems of Gujarat, while associated with the Gujarat Institute of Desert Ecology in Kachchh. His association with The Energy and Resources Institute provided him opportunity to work in different forest ecosystems of the country, from the extreme north in Jammu and Kashmir to the extreme east in Assam and to arid regions of northern Karnataka. While working in these different regions, his focus has revolved around participatory forestry management, addressing livelihoods and poverty issues of forest-dependent communities and indigenous people, and building capacities of grassroots stakeholders for mitigating and adapting to the impacts of climate change.

Dr Silori has published more than 50 research papers and articles in various national and international journals and contributed chapters in a number of books. He is also on the editorial boards of many international journals, including Mountain Research and Development, Journal of Biodiversity, Biodiversity and Conservation, International Journal of Forestry Research, International Journal of Ecology, and Journal of Sustainable Development and World Ecology. He was awarded the Asian Studies in Asia (ASIA) Fellowship in 2008 by the Asian Scholarship Foundation, Bangkok, to work on links between conservation and poverty alleviation in Thailand. In his current position, he is coordinating a regional project on reducing emissions from deforestation and forest degradation (REDD) in the Asia–Pacific region, focusing on seven countries.